DUE DATE	RETURN DATE	DUE DATE	RETURN DATE

STUDIES
IN HISTORY OF
BIOLOGY

STUDIES
IN HISTORY OF
BIOLOGY

5

William Coleman
and Camille Limoges
Editors

The Johns Hopkins University Press

BALTIMORE AND LONDON

The Johns Hopkins University Press, Baltimore, Maryland 21218
The Johns Hopkins Press Ltd., London
Library of Congress Catalog Number 78-647138
ISBN 0-8018-2566-0

Studies in History of Biology is an annual series devoted to the history of biology, geology, and the sciences of man, including anthropology, psychology, and nonclinical aspects of medicine. Extensive articles (up to 35,000 words) will be considered, as will critical review essays, historiographical and methodological essays, and occasional items relating to research materials. Contributors wishing to publish in *Studies in History of Biology* are invited to communicate with either of the editors:

William Coleman, Department of History of Science, University of Wisconsin, Madison, Wisconsin 53706

Camille Limoges, Institut d'Histoire et de Sociopolitique des Sciences, Université de Montréal, C.P. 6128, Montréal H3C 3J7, Québec, Canada

Contents

STUDIES
IN HISTORY OF
BIOLOGY

John Hunter, the Animal Oeconomy, and Late Eighteenth-Century Physiological Discourse

*Stephen J. Cross**

Without a doubt it is by this work [the Museum], above all others, that Hunter has immortalized his name. In his writings we occasionally find an obscurity in the expression of his thoughts, a want of logical accuracy in his reasonings, and an incorrectness in his language, resulting from a deficient education. In this work no such failings are apparent; Nature is here made to be her own expositor, and the treasures she has poured forth come fresh to the mind from the fountains of knowledge, unimpaired by passing through the imperfect medium of language, and unimpeachably proclaiming the genius of him by whose labours they were brought to light.
—Drewry Ottley, *The Works of John Hunter*

Introduction

Attention to John Hunter (1728-1793) in the history of biology and medicine has been by no means lacking. In particular, surgeons in England return constantly to the figure who occupies a special place in their history.

For their helpful criticisms of the 1977 Johns Hopkins University Master's essay upon which this paper is based, I thank William Coleman, Karl Figlio, Donna Haraway, Ludmilla Jordanova, Othmar Keel, and Camille Limoges. A synopsis of my argument was presented to the 15th Annual Meeting of the Joint Atlantic Seminar on the History of Biology and Medicine, University of Toronto, April 1979.
*Department of the History of Science, The Johns Hopkins University. Copyright © 1981 by The Johns Hopkins University Press.

Was not John Hunter, after all, the great anatomist by whose labors their Royal College gained its demonstrative base—the superb Hunterian Museum of Comparative Anatomy? And did not Hunter, too, lay the intellectual foundations that made their art a science? The surgeons have a unique relationship with their history:

> This Hunterian Festival is pre-eminently the occasion in the life of the College when we take the opportunity to draw spiritual and intellectual sustenance from our roots in the past. We refresh ourselves at the seemingly inexhaustible spring of John Hunter's genius. . . . On the foundation of John Hunter's work has been erected the whole superstructure of modern scientific surgery and for that future generations will remember him.[1]

Hunter's legacy has long been institutionalized at the Royal College of Surgeons of England, in the annual Hunterian Orations. Endowed in 1814 "in honour of surgery and in memory of those practitioners by whose labours it has been advanced," the series has frequently taken John Hunter as its subject. Indeed, the majority of historical studies on Hunter convey the particular ideologies of the surgeons—nationalistic, hagiographic, and justificatory: here, it would seem, is less a *histoire sanctionée* than a sanctified history.

In the essay that follows I, too, intend to perform a historical reconstruction. I cannot, however, claim the same privileged relations with its product which the surgeons, members of a corporate body, enjoy organically with theirs. Our interests differ too, and it is not my intent to celebrate; but neither is it, crucially, to assess the significance of Hunter's work for specific developments in the history of the biological and medical sciences. Where John Hunter, biologist, has emerged between the tropes of surgical discourse in recent scholarship, his work has too often been displayed selectively, in order to be either acknowledged as a progressive achievement or regretted as an echo of some dark idea of scientific prehistory. Such "recurrent" readings of John Hunter, which discover problems in their text through the privileged categories of our own biology, diverge as much from the purposes of this paper as do the litanies of the surgeons. My aims and method are neither of the above, but rather to raise the problem of the historical Hunter precisely as a problem, to analyze Hunter's texts through a close reading, to uncover their most important concepts, and to delineate the theoretical space in which those concepts function. On the basis of this conceptual reconstruction I suggest how John Hunter's theories of life, organization, and disease fit into the framework of biological knowledge toward the end of the eighteenth century.

Such a study, then, has been postponed by Hunter scholars in their search for the intellectual influences and prefigurations with which Hunter's story seems to abound; and through their interested readings the surgeon-historians have constituted a patriarch and his kin—both intellectual and institutional—

an authorial projection out of historical time, rather than an historical project in its proper context. My hope is that by the end of the following pages one will possess an exegetical guide to the Hunterian texts; and that in reconstructing a conceptual formation rather than inventorying discoveries or the prescient ideas that contributed so much, supposedly, to scientific modernity, one may gain an interpretative perspective within which it will be possible to "situate" those texts historically. By no means does this paper presume to be a definitive exposition of Hunter's work; its goal is a delineation of the historical specificity of John Hunter's science of the animal oeconomy.[2]

The occasion for this study, focusing as it does upon the work of a single historical figure, is thus the degree to which the subject has remained an underdeveloped area of scholarship in the history of biology. Such ironies as noted above, however, are hardly exceptional for the field. Of more compelling interest than mere exposition is the possibility that a view of John Hunter's physiological problematic—the theoretical context which constitutes the objects of his physiological discourse, and, crucially, defines the limits of it—may enhance the resolution of an emerging general picture of the life sciences in the late eighteenth century.[3] My intention is to realize this potential in two respects, which I shall mention only summarily here, but return to more substantially in the concluding section.

First, my reading of Hunter has been informed by recent findings in the debate over the nature of the transformation of biological knowledge at the end of the eighteenth century. Most prominently Michel Foucault has argued that the discursive strategies, the rules of formation of a discourse and its objects, which for most of the nineteenth and twentieth centuries permit the constitution of a science of life—biology—do not exist in the "archive" of natural knowledge in the classical age—the period from roughly 1650 to 1800.[4] *Life,* Foucault argues, the essential interior processes of a living creature, impossible to represent directly in words, expressed only obliquely in a set of laws of organic structure, vital process, and the necessary relations between form and function, is constituted at the beginning of the nineteenth century as the central object, the organizing concept, of the new discipline of biology. Georges Cuvier's constitutively related laws of the correlation of parts in comparative anatomy, and of the subordination of characters, the basis of the natural method of classification in taxonomy, signify for Foucault the new discursive possibilities of this science. Dramatically, the historian of biology is thus reminded to search for the structure of an historical discourse, and its corollaries—the determinate objects produced, conceptual possibilities, and the silences.

In considering the eighteenth century Foucault concentrates upon natural history, the taxonomical classification of species according to their external perceptible differences analyzed into discrete characters. Recently, however,

William Albury has extended the purview of what Foucault terms *taxonomia,* the discursive ordering of signs in general, to include, in the same period, the physiological classification of vital properties.[5] Albury demonstrates that the analysis of physiological function into the discrete vital actions of organs, fibers or tissues in the classical age, is initiated by a problematic which posits specific anatomically localized vital properties as its essential objects, and which cannot produce the concept of general function, of vital action independent of any particular organ or system. Positivist experimental physiology in the first decades of the nineteenth century, championed in François Magendie's manifesto of 1809, is, for Albury, the product of a fundamental epistemological transformation analogous to Foucault's "epistemic" discontinuity. I have measured my reading of John Hunter against the conclusions of Foucault and Albury on comparative anatomy and physiology respectively. In asking about the historical character of Hunter's project, therefore, I imply a series of broader questions, which may be summarized as follows: what are the contradictory tendencies of this science caught, apparently, within the contours of a profound transformation in natural knowledge?

My second intention flows directly from these questions; for they suggest basic problems of method, and a methodology is chosen specifically to answer certain questions. I have made use of such a method in my analysis of John Hunter's science of the animal oeconomy as might allow that science to be referred to a broader context than physiological discourse itself. In order to establish the historicity of Hunter's project, I have utilized a structural rather than an historical approach; and to situate Hunter in context I have emphasized the rules of a language rather than the real materiality of historical events. Yet the intention of such a synchronic discourse analysis is to provide the basis for adding a social dimension to the picture of natural historical knowledge at the end of the eighteenth century. My chosen method allows these antinomies—structure and history, discursive and non-discursive practices—to be reunited at a level rarely considered by historians of science. This is the level at which, as Karl Figlio has argued, concepts as vectors of ideology flow between one discourse and another.[6] By undertaking what I have termed a conceptual reconstruction, I have attempted to reach this level. Specifically, I ask whether John Hunter's theory of the animal oeconomy should not be viewed as one instance, in this case in the empirical domain of life science, of a general discourse inclusive of eighteenth-century "political oeconomy." Penetrating the surface level of Hunter's texts, the level at which explicit debates appear, I have reconstructed an underlying unity. The material processes of production of this total structured and structuring discourse, no less than the social functions of the more readily accessible ideas, traditions, and disputes between rival schools, must be sought by the sociological history of science. I intend this paper secondly, therefore, to raise the fundamental problem of the "object of the history of the sciences" as a matter of concern for the new social history of science.[7]

This essay is divided into two parts, reflecting not so much an empirical division as a real intellectual tension within John Hunter's project. In the first, the conceptual field of Hunter's comparative anatomy is presented and analyzed; in the second, Hunter's concepts of physiological function, and of the anatomical organization of the individual organism, are subjected to a closer examination. The objective of both parts is ultimately the same; namely, to express the theoretical relation of the concepts of life and organization, rather than simply to specify each concept. The sections differ, however, in the objects in which they situate this relation. My first analysis reveals a comparative anatomy akin in many respects to the science initiated at the beginning of the nineteenth century. Hunter's theorization of his subject, though quite unsystematic, separates his project from the structure of eighteenth-century natural history. In my second analysis, a very different Hunter emerges. Here I portray a physiological project which would seem to fit squarely within both the discursive traditions and the epistemological framework of eighteenth-century physiology. Hunter's science is structured and constrained by this problematic, and as a whole must be understood as the product of a period of profound transformation in the knowledge of living beings; both the beginning and the end of this process are contained within the unity defined in this paper by the published works of John Hunter.[8]

First, I consider Hunter's project of comparative anatomy as it is displayed in his celebrated museum. In two seminal respects physiological determinants entered John Hunter's comparative study of anatomy. On the one hand, as I explain, he defined organ systems, and the series of their anatomical varieties in different species, in terms of their physiological role in the organism. On the other hand, and more profoundly, there is to be found implicit in Hunter's concept of organic structure, and his critique of taxonomy, the teleological notion of the interdependent physiological processes of the animal oeconomy—the concept of life itself. Also in this first section I discuss how the term pathology, for Hunter, denotes the "physiology of disease," or alteration in the conditions of life, and how the concept of death, ironically completing the concept of life in comparative anatomy, enters the theoretical field of nosography.

In the second part of the paper I switch analytic lenses, as it were, and focus more sharply upon Hunter's concept of vital action, or physiological function. One thereby finds that Hunter's physiological theory is constitutive of a project characteristic of eighteenth-century vitalism—namely, the specification of the anatomically localized vital properties of irritability and sensibility.

For Hunter, vital properties constitute the ultimate objectification of the relation between physiology and anatomy, life and organization. In the rest of the section I discuss Hunter's theory of the "internal animal oeconomy," that is, of organization at the level of the internal parts and organ systems, and of life as the continuous processes of production, distribution, and exchange that compose the animal's internal subsistence "oeconomy." The second section concludes with a discussion of the consequences of the problematic of vital properties for Hunter's general theory of the organism, and the interaction of organic parts.

To conclude this essay I bring to the foreground the recent interpretations mentioned above of the general shape and structural changes of the life sciences toward the end of the eighteenth century. While these have guided my interpretations at many stages, I argue that the work of John Hunter is a significant but hitherto overlooked element of the general picture, which latter may indeed be enhanced by including John Hunter within it.

First, however, I present a brief biography, since John Hunter is known better in the annals of medical than of biological science. For this reason also, and because a view of Hunter's work as broad as this has not been attempted elsewhere, I have supplied extensive textual and bibliographic references and commentary.

Part I

The Surgeon Anatomist

John Hunter was born in 1728, youngest of the large family of a grain merchant turned minor landlord in Long Calderwood, Lanarkshire.[9] His formal childhood education was sparse indeed, concluded at the age of thirteen upon leaving the local grammar school at East Kilbride; subsequently, Hunter's lack of letters was to be an integral feature of his renown.[10] When Hunter reached the age of twenty, however, an opportunity to undertake a significant enterprise offered itself. His brother William (1718-1783), ten years his senior, had begun to establish himself in the medical profession in London. In the autumn of 1748 John travelled south to become William's assistant and, should he prove himself able, begin an education in anatomy and surgery. William's success in the foreign capital was to be as great as any a Smollett could have imagined, and John's own eventual rise to fame in surgery, natural science, and metropolitan society, was dependent originally upon it.

William Hunter at the age of thirteen had been sent to the University of Glasgow, where he was to have studied for the Church.[11] However, after completing the curriculum, but without receiving a degree, he became apprenticed to William Cullen (1710-1790), who was then practicing medicine under the Duke of Hamilton's patronage. He stayed three years with Cullen, then in 1739 enrolled at the University of Edinburgh to study anatomy under the famed medical progenitor, Alexander Monro *primus* (1697-1767). The following year, with the hope of gaining a superior training with obstetricians resident in the affluent metropolis, he took the boat south to London.

William never returned to establish a partnership with Cullen as they had planned: he made his fortune in the capital. During the first months he lived with, and studied obstetrics under, William Smellie (1697-1763), who had himself recently arrived from Lanark. It took the gifted and presentable William only a few months, however, to find a suitable local, though equally north British, patron. He became tutor in the household of James Douglas (1675-1742), a well established "man-midwife" and famed anatomist, travelled with his pupil on the continent, and began to make a practice for himself in obstetrics, as a fashion for male midwives developed amongst the wealthy clientele. When in 1745 the Surgeons split with the Barbers, and relaxed their strict regulation of the performance of human dissections, William took advantage of the new legal situation, and in 1746 established a private school of anatomy in a house in the Great Piazza of Covent Garden. It was to help William prepare for the school's third winter anatomy season, and to apprentice himself to his ascendant brother, that John Hunter travelled south in 1748.[12]

Ultimately John Hunter received an excellent education in anatomy and surgery. For eleven years he worked as William's assistant, preparing dissections for the school (and dealing for bodies in the rapidly expanding "resurrectionist" market), making anatomical preparations for William's collection, and assisting him in his research projects. Together they performed dissections for William's superbly illustrated work, *The Anatomy of the Gravid Uterus*. John pursued important investigations of the lymphatic system, and the mode of descent of the testes.[13]

John Hunter evidently showed immediate promise at dissection, since William enrolled him for a summer course in surgery under the famed William Cheselden (1688-1752) at Chelsea Hospital, to which he returned the following summer. When Cheselden had a paralytic stroke in 1751, John is thought to have gone to study at St. Bartholomew's Hospital under Percivall Pott (1714-1788). In 1754 he was admitted to St. George's Hospital as a surgeon's pupil, becoming a House Surgeon in 1756, and ultimately a surgeon member of the Board of Governors there, a post he retained for life. He had already accepted his first patient, although he did not obtain a diploma from the Company of Surgeons until 1768. Thus for eleven years, on the hospital wards and in his brother's dissection room, John Hunter built himself a foundation of surgical and anatomical knowledge.

A break came in 1760 when, in need of recuperation from illness, John obtained a commission as an Army surgeon and set sail with the expedition the following spring to besiege the French garrison on Belle Isle, a small island off the coast of Brittany. After the successful assault, he remained on the island as purveyor to the hospitals, and found time to explore the marine-shore fauna, making dissections of the ovaries of conger eels, and like preparations. The following summer he transferred to the campaign in Portugal, and spent nearly a year in the Alentejo east of Lisbon before returning to London in 1763. As well as bringing back a nucleus collection of two hundred bottled anatomical preparations, he had gained in service clinical experience of the treatment of gun-shot wounds, which provided the subject of a subsequent treatise.[14] He won in addition an army pension, the legal right to practice surgery, and a reprieve from the putrid vapors of the dissection room.

William Hunter had taken on a former pupil William Hewson (1739-1774) as assistant while John was abroad, and no longer required his brother's services. John set up a dispensary in Golden Square, and during the next decade he struggled, and succeeded, at establishing a surgical practice in the competitive metropolitan profession. He gained an entrée in part through surgical dentistry, his interest in which resulted in a first booklength treatise, *The Natural History of the Human Teeth* (1771).[15] When, in 1768, William moved into the impressive building in Great Windmill Street that had been designed for him specially as an anatomy school, John took over the lease on his brother's Jermyn Street property, and began accepting students of his own.[16] He had already, in 1765, acquired land and begun to build a house at Earl's Court, which was at that time in the countryside. This served as his retreat for the pursuit of natural history, and he kept there an ever-growing menagerie of exotic creatures.

The final measure of John Hunter's professional success was taken in 1783 when he purchased a fine large house in Leicester Square. His practice, which was by then attracting a fair portion of the noble clientele, was bringing in several thousand pounds per annum. William Hunter had been obstetrician to Queen Charlotte, and John in his turn became surgeon-extraordinary to the King. In 1790 John was appointed Surgeon-General and Inspector of Regimental Hospitals in the Army. The biographies of the Hunter brothers afford paradigm examples of the potential for upward social mobility that existed for a handful of eighteenth-century medical and surgical practitioners.[17]

John Hunter's reputation gradually spread in scientific circles, too, and during the 1760s and 70s he came to establish connections with such influential naturalists as Joseph Banks (1743-1820), John Ellis (1710-1776), and Daniel Solander (1736-1782). In 1767 both John and William were elected Fellows of the Royal Society. John is said to have been chairman of the inner circle of Royal Society fellows which met in the 1780s to discuss society business at Slaughter's Coffee House in St. Martin's Lane.[18] He submitted twenty-two papers to be read before the Royal Society, the first an anatomical supplement to a paper by John Ellis—who sponsored his Royal Society membership—on an "amphibious bipes" (the Siren), published in the *Philosophical Transactions* of 1766. An additional paper, Hunter's last, entitled "On Extraneous Fossils and Their Relations," was rejected by the Royal Society in 1793 on account of its implication of a non-biblical chronology. A number of these papers gained wider circulation as a collection, *Observations on Certain Parts of the Animal Oeconomy*, published by Hunter himself in 1786 at his house in Leicester Square.[19] Between 1776 and 1782 he read six Croonian Lectures on muscular motion before the Royal Society, and in 1787 won the society's Copley Medal. John Hunter, together with the Royal Society's president Joseph Banks, who owned the world's finest private herbarium and botanical library at his house in Soho Square, no doubt made, through the natural historical sciences, a significant contribution to the Society's revival at the end of the eighteenth century.

Hunter's recognition by the international scientific community is evidenced by a number of honors. In 1781 he was elected Fellow of the Royal Society of Belles Lettres at Gottenburg, and in 1783, a member of both the Société Royale de Médecine and the Académie Royale de Chirurgie in Paris. In 1787, he was elected to the American Philosophical Society in Philadelphia. The Irish College of Surgeons in Dublin, and the Chirurgo-Physical Society of Edinburgh, conferred memberships upon him in 1790 and 1792 respectively. The surgeon anatomist's home *cum* museum was surely a landmark in the international circuit of *savants*. Hunter was visited, for example, by three of the major contributors to the late eighteenth-century revival of anatomy—Peter Camper (1722-1789), J. F. Blumenbach (1752-1840), and S. T. von Sömmerring (1755-1830).[20]

John Hunter's scientific treatises deal with a range of subjects, from experiments on the growth of bones to the natural history of bees; from the nature of "extraneous fossils" to the cause of motion in plants. His papers mainly treat of special anatomical topics—the organ of hearing in fish, for example, the electric organs of *Gymnotus* and the torpedo, the "unusual" stomach of the Gillaroo trout, the secreting crop of the pigeon, the vascular supply of the placenta, the lymphatic system, the vesiculae seminales, and the air-sacs in the bones of birds. He wrote a careful essay on digestion which shows wide knowledge of previous and contemporary literature, and wrote up his experiments and conclusions on animal heat in a well-known paper delivered before the Royal Society in 1775.[21]

Each year during the last two decades of his life Hunter gave a series of lectures on the principles of surgery in which he would expound his theories of life, organization, and the principles of disease, illustrating them abundantly with descriptions of the experiments undertaken for that purpose. Eventually Hunter managed to put his ideas into print, impelled by a characteristic and—given the practice of the times—justified fear of plagiarism by his students, in the lengthy treatise on the blood and inflammation, published posthumously, that paralleled but considerably amplified his lecture course. [22]

Hunter enjoyed a reputation as a scientific autodidact, which he fostered deliberately in a classic struggle against dogmatism in surgery and medicine. As a consequence his genius has frequently been ascribed to an unlettered empiricism, an undistracted fidelity to the Book of Nature; or at least it is believed that "he had no overall theoretical system and he stands in sharp contrast to the armchair theorist and the rationalist system-maker." [23] Both of these views are misleading, and have surely contributed to the general failure to assess John Hunter's work as a characteristic product of its period. While the question of "influences" can lead only to a partial understanding of the historical determinants of Hunter's discourse, and is not raised methodologically in this paper, it is important to challenge the empiricist mythology of his scholarly naïveté. Significantly, those papers in which Hunter makes plentiful citations involve polemics such as priority disputes, or as in the case of the essay on digestion, hypotheses that directly challenge entrenched positions. "I do not mean," Hunter informed his students at the beginning of his series of lectures,

that all I say in this course will be entirely new, and that none of the opinions and observations which I shall deliver are to be found in any publications, for there are many facts too obvious to have been overlooked in the most ignorant days of physic. I intend in this course to include some of the most interesting parts of surgery, principally those subjects on which I have had opportunities of making many new, and I hope useful observations, so as to put our art, in many respects, in a new point of view. Many of my ideas, and the arrangement of my subject, are new, and consequently my terms become in part new. . . . [24]

It suited Hunter's interests to make few acknowledgements in the ensuing lectures. The historian should take into consideration, I suggest, the professional, social, and psychological significance of medical and scientific citation, polemic, and proprietorship in this period.

The scientific enterprise, however, that came to occupy John Hunter above all else during the last thirty years of his life, was what might best be termed imperialism of the internal anatomical space of the animal species, including the invertebrates. In Britain Hunter led the contemporary invasion of the internal anatomy. He was, as a comparative anatomist, seemingly impelled by the principle of plenitude, or continuity, to discover and anatomize every form in the perfectly graded series of organs. As he worked prodigiously at his dissecting table, the collection of two hundred preparations brought back from Portugal grew into the impressive museum of comparative anatomy for which he has been justly acclaimed. He erected a special building on his property in Leicester Square to house the collection, which ultimately contained some fourteen thousand specimens and cost him personally, it is estimated, £77,000.

The story of Hunter's animal collecting and anatomical practice is one of the most picturesque of the period. The established Mr. Hunter became a nag to the nobility, demanding cadavers from their menageries. He maintained contacts around the globe through agents of foreign trading companies, pupils who became naval surgeons, or through such well-placed naturalists as his friend Joseph Banks, to supply him with material for the scalpel.[25] He even sent a surgeon on a Greenland whaler, and dissected several species of *cetacean* (beached in the Thames), mounting their skeletons under a skylight in the yard beside his museum. John followed the example of his wealthy brother in patronizing the frequent sales of cabinets, which record for us one particular enthusiasm of that age. And like William before him in human anatomy, he became the expert comparative anatomist about town, to whom all questions on the subject should be addressed. Everard Home (1756–1832), his brother-in-law and assistant, reckoned that "no new animal was brought to this country which was not shewn to him." At Earl's Court he established the virtual prototype of an experimental research station where he would observe, breed, and perform experiments upon an impressive variety of animals, from lions to bees.[26]

John Hunter's property in Leicester Square has been aptly termed a veritable citadel for the comparative anatomist. The handsome townhouse on the square served as the setting for the social *soirées* of his wife Anne. A second block at the rear on Castle Street (now Charing Cross Road) housed his servants, assistants, and medical students—a reckoning by William Clift in 1792 showed almost fifty persons in Hunter's household—as well as the dissection room and his private printing press. In the space between the buildings he had a lecture theater constructed for his school and for meetings of the Lyceum Medicum Londinense and the Society for the Improvement of Medical and Chirurgical Knowledge, both founded with his close friend Dr. George Fordyce (1736–1802); and above this was a galleried hall to contain the growing collection of spirit-filled flint-glass jars, skeletons, fossils, and dried preparations.[27]

During his last decade at Leicester Square, John Hunter pursued a regular daily routine, leading him into the dissection room at first light, and keeping him up late into the night dictating passages to his amanuensis for the several treatises in progress. In between these engagements with his science he fulfilled those of his profession, the mornings occupied with a surgery at home, the afternoons on the wards at St. George's Hospital. In view of his stupendous labors in comparative anatomy, which manifest themselves to this day in the museum housed since 1806 at the Royal College of Surgeons in Lincolns Inn Fields, it would seem that the Presbyterian work-ethic became imbued in John Hunter with a very special biological curiosity.

In these last years of his life, with his health failing markedly, John Hunter felt increasingly pressed to complete one aspect of his work in particular: the compilation of a catalogue of the huge number of preparations and exhibits in his private collection. This catalogue was intended as more than simply an enumeration and description of individual preparations (and as more than, as Home suggested, an addition to the value of the museum to defray the debts which Hunter's widow might expect to suffer). It was to express the plan, the systematic conceptual framework, by which the collection, at first chaotically shelved, was to be arranged within the cabinets to demonstrate the natural continuities of organic forms. By 1787 the arrangement of the preparations was complete, and the Hunterian cabinet established for the genteel public as yet another source of amusement in Leicester Square; but Hunter completed neither a catalogue nor the major treatise he had hoped to derive from the totality of his labors, and publication of a full description of the Hunterian collection had to wait over forty years until Richard Owen (1804–1892) completed the task as Conservator at the Royal College of Surgeons.[28]

The Renaissance of Comparative Anatomy. In performing its organizational function, the catalogue would necessarily perform another, intellectually more profound. It would provide the link between John Hunter's theories of life and organization, and their demonstration in the dark internal anatomical parts brought out at last into the light of day. For Hunter's museum was, as a later admirer termed it, his "great unwritten book."[29] The aging anatomist became preoccupied with the formidable problems of translation, of express-ing in a written text the grammar of organic structure, which is the set of laws of systematic comparative anatomy. No museum of comparative anatomy existed, or was to exist for some time in Britain, quite like that created by John Hunter. It served as a demonstrative and pedagogic device for the neophyte and expert comparative anatomist alike; and what it demonstrated was radically new. Hunter's museum was far indeed, conceptually, from the *thesauri anatomici* of Frederik Ruysch (1638–1731) at the beginning of the century, whose allegorical expressiveness mediated the ability of the arrange-ments in cabinets to signify anatomical knowledge; and far, practically, from the exhibits of a doctor-collector such as Richard Mead (1673–1754). Even William Hunter's internationally renowned museum at the Great Windmill Street medical school came far short of John's collection in comparative scope; and William's purposes were different in kind from John's intentions in systematic comparative dissection. A discussion of John Hunter's biological work cannot justly begin other than with his museum. [30]

In the following pages I observe the cognitive field of Hunter's systematic comparative anatomy through the medium of the nineteenth-century catalogues of the museum as it was established at the Royal College of Surgeons in London. I focus on particular concepts which constitute this field and define its limits. Anticipating briefly, these are the concepts of: the anatomical series; anatomical as opposed to taxonomic classification; and physiological function as an analogical rather than structural classificatory category. This last concept refers to the comparison of anatomical structures in different animals according to their use in the organism rather than their visible form. Hunter attempted within the various subdivisions of his collection to illustrate what he believed to be the gradual progressive modification of the same organ in different species, and to lay out the variety of structural forms of this functionally defined organ side by side as an anatomical series. [31]

In this respect Hunter's work is innovatively systematic. However, he should not be regarded as a systematic naturalist—or a philosophical zoologist—in the sense of one seeking to provide the anatomical basis for a system of taxonomic classification. His task was anatomical and physiological, but not taxonomic: the classification of organs, rather than of species; the elucidation of the principles of the internal oeconomy, of individual organization, rather than of the natural oeconomy or of the relations between organized forms. Where, however, Hunter made his greatest break with previous anatomical tradition was in the conditioning prior role he accorded to physiological functions in the classification and comparison of anatomical parts. His interest in the structure of a part was inseparable from his interest in its use within the animal machine. In fact Hunter's choice of structures to illustrate anatomical series was made upon functional grounds. The life of the animal, its "natural history," and its "oeconomy," entered the comparison of organization as principles of a higher order than mere structure.

Comparative anatomy had received remarkably little attention for most of the eighteenth century prior to the period of Hunter's working life; until Hunter's time few practitioners matched the range of Malpighi (1628–1694), Swammerdam (1637–1680), or Perrault (1613–1688), working in the previous century. Moreover, until then few studies were genuinely comparative, involved more than a few species, or concerned themselves with internal organs other than the major ones; and almost without exception animal dissections were performed with the traditional medical goal of illuminating that model of all living beings, the human body. Yet looking back on the work of his own generation, and that of the generation immediately prior to himself, Georges Cuvier (1769–1832) could claim that a revolution had been made in zoology through the development of a natural method of classification, and that the basis of this achievement had been a return to the study of comparative anatomy. In Holland Peter Camper, in Paris L.-J.-M. Daubenton (1716-1800) and Vicq-d' Azyr (1748-1794), in Edinburgh Alexander Monro *secundus* (1733-1817), and in London John Hunter had practiced anatomy as if it were truly a science of comparison.[32]

More profoundly, the practice of comparative anatomy in this period was initiated and structured by a conceptual transformation of tremendous importance. Its intellectual goal became the creation of a "philosophical zoology," a set of laws relating the internal structures and the taxonomic relationships of all living forms, the first expression of which being Cuvier's own *Leçons d'anatomie comparée* (1800-1805).[33] Fundamentally this new science of life posited its existence not so much through the production of new concepts, though we recognize it through such, as by its nature as a discourse and a material practice—both characterized by specific rules that allowed the life of an organism to be expressed—or as an epistemological framework or theoretical problematic in which certain concepts became thinkable. Indeed, the very notion of "life," the object of biology, can be said to have come into being with the expression of the character and identity of an organism not in terms of the elements of external appearance—of an algebra of signs as in eighteenth-century taxonomy—but in terms of the mysterious internal processes of the organism's vital being expressed in concepts that could only be generated discursively. Here is the significance of statements such as Cuvier's principle of the functional correlation of parts, and of the practice constitutive of the new science, comparative dissection, that sought to uncover the internal structure of life. In the first part of this essay I explore the question of the relation of John Hunter's discourse to this epistemological break with the natural history of the eighteenth century.

In the preface (1837) to his edition of Hunter's *Observations on Certain Parts of the Animal Oeconomy*, Richard Owen made his own assessment of the issue. He suggested that

> the historian of the natural sciences had just and sufficient grounds for regarding Hunter as the first of the Moderns who treated of the organs of the animal body under their most general relations, and who pointed out the anatomical conditions which were characteristic of great groups of classes of animals; as one, in short, throughout whose work we meet with general propositions in comparative anatomy, the like of which exist not in the writings of any of his contemporaries or predecessors, save in those of Aristotle.[34]

My ensuing discussion is designed to assess such an historical judgment (which is not confined to English anatomists of Owen's generation), and to situate John Hunter in the history of the sciences.

Anatomical Series

As a Method of Arranging Exhibits. John Hunter divided his anatomical collection into three main parts.[35] The first contained structures illustrative of the functions that minister to the necessities of the individual (the general functions of digestion, circulation, locomotion, etc.); the second, structures that provide for the continuance of the species, that is, the organs and products of generation; and the third, pathological preparations. His exhibits in pathology consisted of 625 dry preparations and 1,084 preparations preserved in spirit. The latter preparations in spirit were subdivided under three heads: those illustrative of the general actions of disease and of restoration (for example, inflammations); the anatomical effects of specific diseases (scrofula, cancer, syphilis, etc.); and specimens of diseased parts, arranged according to their anatomical locality (the esophagus, stomach, spleen, etc.).

The museum contained, in addition to these three main groups of exhibits, 617 dry preparations of anatomical structures; 965 preparations of human and comparative osteology; an exhibition of monsters;[36] 1,743 specimens of natural history in spirit (whole animals, including invertebrate preparations);[37] stuffed animals; about 1,000 dry specimens of insects, shells, "zoophytes," etc.; calculi and other concretions such as bladder stones; and a collection of 3,709 fossils.[38] In November 1799, when the museum was handed over to the custody of the Company of Surgeons it contained some 13,682 specimens, and represented about 500 different species of animal.

The first two main divisions of exhibits in John Hunter's museum were originally separated by the pathology division, but subsequently became catalogued together as the "Physiological Series of Comparative Anatomy."[39] They demonstrated normal anatomical structures, and together contained 3,745 preparations in spirit, being the largest division of the original museum. It is the concept of "series" that I propose now to discuss.

Richard Owen, in his preface to the first catalogue of the Physiological Series of 1833, described this division of the museum as follows:

The department of the Hunterian Collection to which the present Catalogue relates, is devoted to the illustration of the highest branch of Natural History —the science of life itself.

It consists of dissections of plants and animals in which the *structures subservient to the different functions* are skillfully and intelligibly displayed.

These structures are taken from every class of organized matter, and arranged in *Series,* according to the function, in the order of their complexity, beginning with the simplest form, and exhibiting the successive gradations of organization to the most complex.[40]

Hunter's fundamental organizing principle is the concept of the physiologically defined anatomical series. At a first level this principle is used to
generate subdivisions according to general functions. The term series refers in
this sense to a set of examples of anatomical structures that have a particular
physiological use in the animal body. Thus the first division of the
Physiological Series, exhibiting organs in plants and animals for the "special
purposes of the individual," is subdivided under nine functional heads:
organs of motion, digestion, nutrition, circulation, respiration, excretion,
organs that bring the individual into relation with the external world (the
nervous system and organs of sense), connective and adipose tissue, etc., and
lastly, peculiar organs that serve special uses in particular species. Each sub-
division is itself then divided into "series" illustrating the various structures
that perform component actions of the general function. [41]

Richard Owen's description makes clear, however, that the concept of
series serves a more precise role than the above. By the concept of series John
Hunter referred not simply to a set of related anatomical parts, but to
"successive gradations of organization" within each (functionally defined)
class of organs. "The Collection consists of examples illustrative of the various
structures of similar organs in different plants and animals; so arranged that
each set of organs forms a distinct series, or link of a chain of gradations of
structure, from the most simple to the most complex." [42] The concept of the
series postulates an *order* of structures within each set.

Many of the series in the nineteenth-century catalogues devised for the
museum at the Royal College of Surgeons are arranged internally according to
a supposed order of complexity of the organs. The structure of the intestines,
for example, is illustrated by a series which begins with those in the Annulose
animals, proceeding to Molluscs, Fishes, Reptiles, Birds, and finally Mammals. [43] Preparations, therefore, in that part of the museum devoted to
normal anatomical structures, were classified according to function, and
arranged according to the order of structural gradation. This was the
heuristic organizational role of the concept of physiologically defined
anatomical series.

As a Biological Principle. Logically, the formation of a series is the
inevitable result of comparison involving more than a pair of objects related
by category. To limit one's perception of Hunter's concept of the physiological series to its role as an organizational principle, however, would be to
miss its more profound meaning at the biological level; one would overlook
the sense in which the museum was John Hunter's "great unwritten book,"
the language of which obeyed the grammatical rules of organic nature itself,
rules expressed by the concepts of perfection, complexity, series, and
organization. In fact Hunter believed that the order of an anatomical series

reproduces the order of nature; examination of his writings on comparative anatomy will demonstrate this amply.

When treating of a particular organ system, Hunter projects as the goal of comparative dissection the discovery of every variation in form of that particular organ through the whole class of animals which possess it. Thus, in the case of the digestive system of birds, "every gradation of the stomach," Hunter asserts, "is to be found among them, from the true gizzard, which is one extreme, to the mere membranous stomach, which is the other." He means "every gradation" in the literal sense of a perfect series of forms, for, in consequence of this, "it must be as difficult to determine the exact limits of the two different modes of construction to which the names of gizzard and stomach specifically belong, as, in any other case, to distinguish proximate steps in the slow and imperceptible gradations of Nature."[44]

Significantly, Hunter's discussion of gizzards occurs within the context of one of his several papers on unusual, or seemingly anomalous anatomical parts; in this instance, "Observations on the Gillaroo-Trout, Commonly Called in Ireland the Gizzard-Trout."[45] His purpose in this paper is "to inquire whether [the digestive organ's] resemblance to a gizzard be sufficiently strong to render the term of gizzard-trout a proper appellation, and what place its stomach ought to hold among the corresponding organs of other animals." After comparison with the stomachs of other fish he concludes: "So far as we are led to determine by analogy, we must not consider the stomach of this fish as a gizzard, but as a true stomach."[46] His purpose therefore is to demystify the anomalousness of an anatomical species, and his method is to insert the part into its respective place in the graded series of forms of its class. The concept of the anatomical series thus functions as an a priori biological principle in the conceptual field, just as it functions as an organizing principle on the shelves of the museum.

Occasionally the demystifying function of the concept of series strengthens, in Hunter's anatomical theorizing, to the degree that the concept becomes a kind of *post factum* predictive principle. Such is the case with a Hunterian anatomical favorite, the organ of hearing in fish.[47] Hunter submitted a short paper to the Royal Society on this part, typically to assess his claims to priority of discovery against those of various continental anatomists (his contribution consisted mainly in clarifying details of the external opening); but he is also concerned again to imply that the anatomist should not regard this recently discovered organ as an anatomical anomaly. On the contrary, he himself is "inclined to consider whatever is uncommon in the structure of this organ in fishes as only a link in the chain of varieties displayed in its formation in different animals, descending from the most perfect to the most imperfect, in a regular progress."[48] What the varieties have in common is their use in the sense of hearing, and this purpose defines the organ discovered. Indeed the anatomist might have been led to search for the particular structural variety of the organ that provides fish with their sense of hearing, supposing that fish can hear. The concept of physiological series would then function as a predictive principle.[49] Hunter's concept of the series has therefore an analytic consequence—a graded scale—and is also the basis for synthetic arrangement, since it provides the a priori specification of relations between particular forms of an organ in a different species.

Taxonomy and the Great Chain. The use that John Hunter makes of the notions of perfection and complexity to evaluate the order of structures in a series is, as the example of the stomachs of fish demonstrates, somewhat ambiguous. Generally he is content to intuit that a heart with two cavities, for example, is more nearly perfect than a single chambered heart. Or at least, to argue that the more complex heart must have more complex relations with other parts, and that the more complex system, as such, is to be taken as the more nearly perfect. In his own terms, he describes the structurally most diversified system as the most "complete," implying a model of completeness—the human body, no doubt: "The complete system is always to be considered as the most perfect, although it may belong in other respects to a more imperfect order of animals."[50] The full meaning of this notion, which suggests a principle of the physiological division of labor, will become apparent in the discussion of Hunter's theory of function later in this essay. Generally, Hunter seems to follow the anthropocentric convention of the chain of being concept in speaking of a descending scale of perfection of whole animals, but as a comparative anatomist he speaks of an ascending scale of complexity in organ series.[51] None of these problems is given sustained attention; but Hunter has much to say about two important consequences of his concept of anatomical series, first for the notion of a single linear scale of beings, and second for the classification of whole animals into the higher order taxonomic groups.

At the beginning of an essay entitled "Observations on Natural History," Hunter considers the general problems of classification.[52] "In Natural Things," he observes, "nothing stands alone"; every natural production is related to some other by virtue of its possessing corresponding parts. In fact, concerning organisms, "there is no whole but possesses several properties that are peculiar to some others." For example, "the four-stomached animals have somewhat similar teeth and cloven feet." The problem for classification is that these resemblances can "spin out ad infinitum." In the limit, we would say, any class would break down since any species will have some resemblance to some other. The solution to this problem is that in classifying organisms, "it is only the great distinguishing parts that should be arranged." The heart, for example, would be one such part; and organisms might then be further classified according to the heart's structure, for instance, by the number of its cavities: *Monocoilia, Dicoilia, Tricoilia, Tetracoilia.* Once animals have been thus classified, they can be arranged in a single linear series as the part increases in complexity, as above.

John Hunter, however, seemingly from the vantage point of his immense experience as a comparative anatomist, believed that any such classificatory scheme, and any such scale of beings, is bound, in the nature of things, to be artificial. And the choice of a distinguishing organ system is bound to be arbitrary. As one organ system increases in perfection, another at least will decrease. Animals, in short, are not "regularly progressive in every part from the most imperfect to the most perfect." For example, considering commonly recognized taxonomic groups, he observes:

Fish are an inferior order of beings to Fowl; yet Fish have teeth, a property belonging to the highest order; therefore Fish in this part step over Fowl. The Amphibia, which are between Fish and Fowl, in this respect resemble both, having both teeth and bill; but then they are most like the Fowl in the vital parts, whilst they have teeth like the Fish. [53]

Hunter is even prepared to admit that contrary to the traditional axiom, "The human body is not a standard in every part for Comparative Anatomy; for though the brain may be a standard, yet the teeth are not." Human teeth come in between those of the herbivorous and those of the carnivorous animals, beings of the "mixed kind." (This is an example in which the concept of perfection is again ambiguous.)[54]

Frequently John Hunter does have a use for the traditional concept of a single linear scale of beings. For example, in the case of the difference in arrangement of the heart-lung organ system in "amphibia" (in this case, turtles) and in fish, he observes that although most anatomical differences between these classes are great, nature, "always proceeding by the nicest gradations," has provided intermediary animals which "partake so much of the two classes, that they gently lead us on from one to the other"—the siren, for example, whose aorta divides like the arteries in the gills of fish, but whose lungs are supplied like the amphibia. He summarizes this scale of organisms: "Thus the gradation is formed from perfect lungs, first to perfect lungs and imperfect gills, then to perfect lungs and perfect gills, till at last we have no lungs, but simply perfect gills, as in fish." [55] Behind his familiar use of the traditional concept, nevertheless, Hunter has a critical awareness of the necessarily artificial character of any single linear order of species based upon just a single organ system. His rigorous use of the concept of a scale of forms is restricted to the level of anatomical parts, without application to the level of whole organisms. [56]

Hunter does make use of familiar taxonomic notions of the "great classes" of animals, mainly those of everyday language, such as Fowl, Reptile or Fish. But he explicitly argues that the lesser divisions of class, tribe and genus are "arbitrary," species alone being an "absolute" unit of classification.[57] In consequence, Hunter offers in his essay on natural history a number of alternative schemes of classification each based upon either a single organ system, or upon a single physiological characteristic: hearts (according to the number of cavities); breathing organs (lungs to gills); brains; parts of generation; mode of coition; body temperature; size of body; element frequented.[58] Elsewhere he suggests classification by stomachs as the most plausible.[59]

Just as Hunter restricts the concept of graded scale to the anatomical level— series of anatomical parts rather than the chain of beings—so too he restricts classification to the arrangement and description of organs. He is concerned with anatomical, physiological and pathological classifications, but not with ordering whole animals into taxonomic groupings. He criticizes Linnaeus who, taking supposedly characteristic parts only, the mammae, attempted to subdivide the different mammals according to the situation of those parts on the surface of the body. Rather, he argues, Linnaeus should have classified the nipples (anatomical parts) according to their situation on the body, not the whole animal. The latter classificatory scheme, Hunter believes, is necessarily artificial.[60]

Hunter wrote but a single treatise of descriptive zoology; it was on the strange "nondescript" animals of Australia.[61] His comments are devoted almost wholly to analogies with more familiar species. The kangaroo, for example, has hair the color of that of the wild rabbit; its foreteeth in the upper jaw agree in number with those of the hog, while the teeth of the lower jaw agree in number with those of the rat; the middle toe of its hindfoot looks something like the long toe of the ostrich; and so on. Hunter had hoped to be "able to form an opinion of the particular tribe of animals already known to which the kangaroo should belong," but found himself unable to do so. The problem for classification is that although "great classes" such as quadrupeds, birds, fish, etc., are "so well marked that even the whole [animal] is justly placed in the same class . . . when we are subdividing these great classes into their different tribes, genera and species, then we find a mixture of properties, some species of one tribe partaking of similar properties with a species of another tribe."[62]

Hunter makes routine use, then, of traditional concepts of the "great classes," and is often prepared to assign these classes relative places in a great chain of beings; but he argues at the same time that lesser taxonomic groupings based on a single organ system as the distinguishing sign must necessarily be artificial. When in his essay on natural history Hunter argues for the division of the vertebrate species into the "great classes," he gives an *extended list* of the anatomical (and physiological) characteristics of species within each class, distinguishing between several that are "essential," and those that are "circumstantial."[63] Hunter's criticism of taxonomy derives, therefore, from his perception of anatomical complexity, or rather—and I return to this crucial distinction below—from his concept of organic structure. With this concept one has the key to unlock John Hunter's whole problematic; but first, it is necessary to highlight the perspective on his functionalism.[64]

Function and Life

In discussing John Hunter's concept of the anatomical series, and its consequences for taxonomy, I have repeatedly referred to the determinant role of the concept of function. I propose now to focus on his notion of the relation between physiology and comparative anatomy.

Analogy. Hunter defines anatomical organs physiologically. They are structures specially formed to execute particular functions, to answer specific purposes in the organism. This functional definition is implicit in a metaphor that Hunter uses to explain affinities between different classes of objects: "A watch is a species of instrument for dividing time, whatever materials it may be composed of; and though a watch is indivisible, yet it has relationships, such as to a clock or a dial."[65] A watch has no (or, in being compared with a clock, few) structural parts in common with the other instruments. But since they all have the same use, there exists a functional genus of "dividers of time." Again, in the context of a discussion on teratology, Hunter argues that the structure of monsters may explain something in the physiology of normal animals "just as the 'weight' in a clock might explain the use of the 'spring' in a watch, etc."[66] The analogy is between functions rather than structures.

This metaphor aptly expresses one element of the conceptual basis of John Hunter's comparative anatomy. The conceptual space in which anatomical structures of different organisms are compared (or the physical space in which the bottled preparations of anatomical parts are exhibited) is the "anatomical series"; and a series is defined under the head of the common function of the various structures, analogically. Consider the series of organs for performing the function of respiration:

Every part of the animal so exposed to the air as for the blood to be affected by it, in such manner as to support life, may be called Lungs, or Respiratory Organ; but what is commonly understood as such, is an apparatus formed for that purpose, as a distinct part of the animal. . . . Where there is such an apparatus, we find it admits of forms fitted for the different modes of respiration; yet all are included in the term *Branchiae* or Gills, and *Pulmones,* or Lungs.[67]

Gills and lungs are adapted, Hunter contends, to their respective modes of respiratory exchange, and are structurally quite different from each other. Their general function, however, is the same. Similarly, although in every class of animal the heart is "the engine employed to throw the blood to those parts into which the arteries conduct it," the form of the heart in different species is extremely varied, answering to a particular "immediate use":[68] "The heart varies in its structure in different orders of animals, principally with respect to the number of cavities and their communications with each other, yet in all nearly the same purpose is answered."[69]

Physiology and comparative anatomy are therefore two aspects of a single mode of investigation for John Hunter. In the physiologically defined concept of the anatomical series, structure and function are constitutively related. Indeed, Hunter believes that comparison in anatomy has a physiological goal: to "make a series of anatomical preparations, perfectly classed according to their uses only."[70] He is not interested in constructing these series simply to demonstrate morphology. Comparison of anatomical structures gains its greatest value as evidence of the actions of the living body. Hence, as a surgeon: "The human body is what I mean chiefly to treat of; but I shall often find it necessary to illustrate some of the propositions which I shall lay down from animals of an inferior order . . . in order to show from many varieties of structure, and from many different considerations, what are the uses of the same parts in man."[71]

At first sight such intentions seem merely to confirm that John Hunter, like most anatomists who investigated the animal oeconomy in the classical period, viewed the bodily frame and its organic parts as an *anatomia animata,* to use Albrecht von Haller's phrase, a living structure of which each anatomical element must have a specific physiological function. Within this anatomical tradition—which had classical roots in Aristotle and Galen, reemerging in the middle of the seventeenth century with Harvey, Willis, and the natural theologians, and continuing in England into the early decades of the nineteenth century until physiology as a discipline was freed from its subsumption under anatomy and surgery—every structure was ascribed a function, and hence discovery in anatomy meant simultaneously the elucidation of physiology. John Hunter's work was unquestionably informed by this conceptual formation, and in a very precise manner to which I devote the second half of this essay. The anatomical dictates of Hunter's physiology separate it, as a discrete problematic, from the physiological science of the nineteenth century.[72]

Nevertheless, to satisfy oneself with this initial perception is to miss the real importance of Hunter's concept of the functionally defined anatomical series within his project in comparative anatomy. For not only must every structure have a function, but the same general function may be performed in different creatures by utterly different structures. While the anatomical problematic discussed above prescribed that to define an organ or part is to disclose a specific function, an anatomical series presupposes the opposite, in that according to previously defined functions it is possible to specify the identity of different structures. In other words, as I demonstrated with the example of the series of lungs and gills, the identity of organs cannot be signified by their visible form; no longer can a lung be identified and *classified* as a visible, tangible presence, but only as the instrument of an ineffable non-existent—as an "organ of breathing." A concept must now serve where a name, a character, had before, and only a whole theory of functions, a discourse separate from, indeed constitutive of its objects, will serve to organize the forms that had previously, as it were, organized themselves in their immediate being. Here is the real importance of John Hunter's concept of the functionally defined anatomical series; for it presupposes the fundamental underlying unities of biology—general functions—and it uses them explicitly to arrange and classify forms.[73]

In replacing visible structure with *function* as the most basic taxonomic protocol, therefore, Hunter has begun to displace the principal epistemological a priori of eighteenth-century natural history. His frequent complaint is that not lack of anatomical knowledge, but a "deficiency in the knowledge of their oeconomy" has prevented the classification of certain animals. For example, on the problem of classifying marsupials, Hunter comments that the preserved specimens sent home from New South Wales "may partly explain their connection with those related to them, so as in some measure to establish their place in nature, but they cannot do it entirely; they only give us the form and construction, but leave us in other respects to conjecture, many of them requiring further observation relative to their oeconomy."[74] This privileging of physiology in taxonomy, however, does not in itself demonstrate that Hunter's project was truly biological; the arrangement of organs according to functions does not lead inevitably to the comparison of organic structure, of organized beings, according to the inner, ineffable matrix of functional relations that is the object of biology. To explore the question of John Hunter's situation in the formation of biology (as defined in the introduction to this essay), it is necessary to elaborate upon his theorization of the relations between functions—or the functional relations between structures.

Adaptation. The concept of function inscribes within itself a second teleological notion—the concept of adaptation.[75] A structure must be adapted to the performance of its task within the body. The concept of adaptation has, in John Hunter's biological problematic, a three-fold reference.

First, an organism is adapted to its external environment. Hunter argues, for example, that the whole animal is adapted, for the purposes of motion, to the medium in which it lives. The whale, for example, is not a member of the class of fishes (and its "oeconomy" bears no relation to that of the latter), but, like species of fish, its bodily form, skeleton, and musculature are adapted to progressive motion through water. Organs of "transaction" with the environment in any species are likewise adapted to external conditions. For example, the jaws, teeth, oesophagus and stomach are all adapted to the diet of the animal. To take the example of the stomachs of different species of trout again:

The two extremes of true gizzard and membranous stomach are easily defined; but they run so into each other, that the end of one and the beginning of the other is quite imperceptible. Similar gradations are observable in the food: the kinds suited to the two extremes mixing together in different proportions adapted to the intermediate states of stomach. . . .
The gradation from gizzard to stomach is made by the muscular sides becoming weaker and weaker, and the food keeps pace with this change, varying gradually from vegetable to animal. In one point of view, therefore, food may be considered as a first principle, with respect to which the digestive organs with their appendages act but as secondary parts, being adapted to and determined by the food as the primary object.[76]

Lungs are likewise adapted to their particular mode of respiratory exchange.

Second, organs within the body are mutually adapted, structurally, to each other. For Hunter the most important instance of this in the animal oeconomy is the relation between heart and lung. Indeed, one ought really, he contends, to consider these two organs as a single organ system, both being essential parts of the circulation:

Without a collecting and a motion of the nutritive juices, I can conceive there can be no respiratory organ; for I find that the different circulations in the different orders of animals, as far as I know, are so connected with different kinds of lungs, as for either system not to be intelligible alone. For . . . the different circulations cannot be described without including the respiratory apparatus, as this makes a portion of the circulation.[77]

The different modes of construction of the vascular system in different animals results from the particular relation between the structure of the heart and the pulmonary circulation. Likewise, to take a second example, as stomachs become more complicated in different species, parts "preparatory and subservient" to digestion become proportionally more complicated.[78]

Amongst the manuscripts published in *Essays* is a fragment entitled "Treatise on Animals," intended by Hunter as the outline sketch of the anatomical part of his planned multi-volume work on Comparative Anatomy and Physiology.[79] It begins with programmatic statements on the correct order of "Progress for the Study of Anatomy." Anatomical parts, Hunter asserts, should be investigated in the order of their increasing connection with other parts

for, however an animal body may seem to be compounded, all its particular parts having a dependence upon one another; yet in an anatomical sense they are more or less distinct, so as to admit of distinct examination. This connexion or interdependence is not in an equal degree, some parts being immediately connected with a greater variety of other parts than others are. The order or degree of connexion is progressive, and will not admit of being reversed. . . . Therefore the parts that have the least degree of connexion should be first considered; because when perfectly understood, those which have an immediate connexion with them will be more easily understood.[80]

His prescribed order is: bones; cartilages and ligaments; muscles; viscera; blood vessels; nerves.[81] Hunter proposes, therefore, a hierarchy of structural relations within the organism.

To fully explicate this concept, however, it will be apparent that one must introduce another dimension to Hunter's notion of organic structure. This is the concept of functional dependency. The hierarchical order of organ-systems is defined, implicitly, in terms of the order of importance of the respective organs for the support of the life of the organism. The vascular system is thus the most important, since every part depends upon a supply of blood. It forms the basis of the "internal animal oeconomy," or system of essential general functions (nutrition, respiration, etc.).[82] The skeleton and muscular system is on the other hand virtually inconsequential from the point of view of the immediate, necessary vital actions.

Hunter's functionalist problematic contains, therefore, a third concept of adaptation: the mutual adaptation, or interdependency, of the functions of the organ systems necessary for life. Indeed, organs and parts are only structurally adapted insofar as that enables them to act harmoniously and hence perform the functions of the integrated animal oeconomy. The determinant of organic structure is life itself.

Action is necessary for the various purposes for which the animal is intended, and if one species of action takes place, it brings the whole into action, as all the parts and actions of the animal body are dependent on one another. If the heart acts, the lungs must fulfil their part; the stomach must digest, the other parts subservient to this organ must be put into motion, and the secretory organs, nerves, and voluntary muscles. The whole is thus set into motion to produce some ultimate effect which appears to be the propagation of the species, for preservation (of the individual) cannot be called the ultimate effect.[83]

My characterization of Hunter's implicit notion of the hierarchical functional interdependency of parts no doubt recalls the famous teleological anatomical rule expounded by Georges Cuvier, the principle of the "correlation of parts." Cuvier's law was elaborated expressly as a principle of anatomy, specifying the relations between organs that are possible on physiological grounds, and that will therefore ensure that harmony of functions necessary as the condition of existence of the organism.[84] No such explicit rule is to be found in John Hunter's texts. My reconstruction, however, has sought to draw out the implications of his functionalist comparative anatomy, and its soundness may be measured by referring to my earlier discussion of Hunter's critique of taxonomy.

I suggested that the reason for Hunter's rejection of a linear chain of beings, or of any classificatory scheme based upon a single dominant organ system, lay in his notion of the complexity of anatomical organization. I implied that his perception of the problem of complexity, of the contradictions of series themselves, was perhaps gained inductively from the tremendous empirical extent of his dissection. The explanation is offered that once the internal space of the organism had been exposed to the anatomist's gaze, organic complexity, immediately apparent, subverted the classificatory signification of surface marks. Upon the dissecting table, organisms were evidently not "regularly progressive in every part." "Birds, for example, have the four-chambered heart and perfect respiration of mammals, but have the oviparous generation, and the general plan of structure of Reptiles."[85] In such an account, John Hunter is made out to be a systematic natural historian whose disclosure of the internal characters of a species simply adds new signs (besides those of the surface) to the grid upon which interspecific relations are gradually summed. Owen suggests just this:

We have thus traced Hunter in his character as a systematic zoologist through a series of attempts at the arrangement of the animal kingdom, in which, like the indefatigable Adanson in a sister science, considering each organ by itself, he formed, by pursuing its various modifications, a series of groups characterized by that organ alone; and doing the same for another organ and another, thus constructed a collection of systems of arrangement, each artificial, because each was founded upon the variations of a single assumed organ.[86]

Such an account, however, fails to recognize the nature and importance of Hunter's concept of organic structure, a concept that imposes itself, moreover, upon the empirical field. For it is precisely this concept, above all others, that places an intraversable distance between the botanical natural historian Michel Adanson (1727-1806) and the biologist John Hunter. According to the problematic of *biology,* which emerges in Hunter in a manner that my ultimate purpose in the first part of this essay is to explain, organic structures are referred not simply to a series or table of characters that, as a system of signs, specifies in language classificatory relations, but to the interdependent physiological functions that guarantee life itself. The identity of an animal class is conditioned by the inward essential processes of its oeconomy.[87] Consider the example of whales: "Although inhabitants of the waters they belong to the same class as quadrupeds, breathing air, being furnished with lungs and all the other parts peculiar to the oeconomy of that class, and having warm blood; for we may make this general remark, that in the different classes of animals there is never any mixture of those parts which are essential to life, nor in their different modes of sensation."[88] The oeconomy of life itself defines structural possibilities and taxonomic classes. So, when Hunter offers lists of the organ systems that characterize the "great classes" (see above), he introduces the concept of organic structure into taxonomy. But more significantly, by labelling certain parts, their situation, and relations "essential" ("a heart made of four cavities . . . the lungs confined to a proper cavity"), and by introducing physiological characteristics ("respiration quick"), Hunter introduces the dimension of life.

I suggest, then, that John Hunter's discourse on the comparative structures and functions of organized beings is capable of generating and sustaining the conceptual problematic of biology; although it is so unconsciously, as it were, through confronting specific contradictions in the writing of a text in the material form of anatomical preparations, rather than as the result of a deliberate rewriting of the principles of philosophical zoology. While organic structure disrupts the order of representation of taxonomic relations in natural history, or as Michel Foucault has put it, "intervenes between the articulating structures and the designating characters—creating between them a profound, interior and essential space," biology refutes the very "conditions of possibility" of natural history as it is found throughout the classical period.[89] Already in Hunter it is no longer possible for the relations of form within and between species to be represented in the order of designating characters—cavities of the heart, for example; the essence of a species must be uncovered in the essential order of its parts, an order that can be given expression only discursively in the generalizations of a science of the organism, of life.

Although this discipline is in reality a specific mode of production of discourse, it has the appearance of phenomena in the realm of scientific competition and exchange—debates, hypotheses, and institutional divergencies and discontinuities at all levels. A major debate made possible in the nineteenth century by the constitution of biology concerned the tension between morphological and functional approaches in comparative anatomy. Of the many editorial comments in the nineteenth-century edition of Hunter's texts, one in particular concerns this debate, and represents the small number of such comments which share the same "historical a priori" with the author.

Functional considerations dominated John Hunter's interest in anatomy, as I have shown. Indeed, he takes the great seventeenth-century anatomist Swammerdam to task for having investigated the anatomy of the bee too minutely. Knowledge of structural minutiae is not only unnecessary, Hunter complains, but "improper" if knowledge of the oeconomy of the bee (its physiology and natural history) is lacking. Richard Owen takes issue with Hunter here in a footnote:

> If the objects of the comparative anatomist were limited to the elucidation of the function of the organs he dissected, there might then, perhaps, be some reason in the animadversions in the text; but his researches have a still higher aim, viz. to trace the general plan which pervades the construction of animals amidst the various modifications to which each organ is subject in reference to particular functions; and the study of organic homologies requires that attention be paid to the minutest particulars, independently of considerations as to the uses in the oeconomy to which they may be subservient.[90]

Owen is here opposing the basic tenets of transcendental morphology to Hunter's functionalist bias. Yet both of these theoretical perceptions of the necessary principles of organization of living beings pivot around the same notion of organic structure, which made its appearance at the end of the eighteenth century. For the biologist it is impossible to compare anatomical structures according to their immediate, visible configuration; their identities and differences can be defined only in terms of a conceptual structure given a priori—a unity of plan, or a general division of physiological functions. Owen's homologies and Hunter's analogies—neither are thinkable without some *biological* principle of comparison. Hence, while I have stressed Hunter's functionalism in the foregoing argument, such a position was only one of several determined by the underlying structure of biological discourse, historically consequential though it undoubtedly was.[91] It remains for me now to compare in greater detail Hunter's implicit biology with the explicit formulation of a rational science of comparative anatomy at the beginning of the nineteenth century.

The Cuvier Transformation. My discussion of the role of the concept of physiological function in John Hunter's project of anatomical comparison has moved from the concept of the physiologically defined anatomical series, to the more profound notion of the internal organization of the animal oeconomy and its dependent processes. I have already indicated that there are striking continuities between the work of John Hunter and Georges Cuvier, in whose texts the new science of life first gained systematic expression. From this viewpoint a more detailed comparison between the projects of Hunter and Cuvier becomes possible; indeed, one is led to suggest that if a "Cuvier transformation" may be said to have occurred in the knowledge of living beings around the turn of the century, then the work of John Hunter should be said also to reside within it.[92] In the first of Cuvier's widely influential *Leçons d'anatomie comparée* (1800), entitled, "Considérations préliminaires sur l'économie animale," Cuvier outlines all the basic elements of the new problematic except a definitive assertion of the order of subordination of parts, and it is this text that best affords comparison with Hunter's theory of the animal oeconomy.

Cuvier begins with a discussion of the general functions of a living organism and the organs by which they are performed. He places major emphasis on the general processes serving the functions of nutrition, which he defines generally as the processes of transformation of fluids. It is here that Cuvier makes his famous analogy of the living body as a hearth (*foyer*) through which matter flows forming and escaping from the organization of life. Cuvier regards this general motion common to all parts of the body, a whirlpool determined by the vital force, as the essence of life. I discuss John Hunter's theory of internal organization and the general processes of circulation and nutrition in the second part of this paper, and shall postpone comparison of it with Cuvier's theory until then. Here I am concerned with the third through fifth sections of Cuvier's anatomical lesson.

Cuvier observes that organs that perform the same function in different animals may have totally different structures, offering the respiratory structures as an example.[93] He outlines the principal variations in the organ systems, defining these physiologically in a manner similar to Hunter's comparison of structures by means of functionally defined anatomical series. Cuvier then proceeds to a proposal and explication of his famous principles of the conditions of existence of a living being—the correlation and coexistence of parts. Not any combination of organs is possible in nature, he argues, only those which allow harmony of function, since the functions are mutually dependent upon each other. Cuvier takes the mode of respiration in relation to the other functional systems as an example, recalling the emphasis that John Hunter puts repeatedly on the interdependency between the respiratory and circulatory systems, and their determination by the natural history, or external conditions of existence of the animal.[94] Like Hunter too, Cuvier argues that the different parts of a single system, the alimentary system for example, are subject to the same law of correlation.[95] John Hunter distinguished the essential from circumstantial parts; Cuvier recasts this in terms of his explicit law, distinguishing between the small number of combinations of the important organs possible, and the seemingly unlimited variety of accessory parts that do not affect the harmony of the whole. Finally, Cuvier criticizes as does Hunter the naturalists' notion of a single scale of beings. He argues that while it is possible to construct uniform series of particular organs, at least through the species of a single class, the order of degradation is not uniform for all parts in any series of whole animals. In order to arrange species according to each organ taken separately, it would be necessary to calculate the position of each species from the combination of its positions in every series of organ, which is hardly possible. Cuvier implicitly criticizes the methods of classification of the naturalists here. In all of this, he does not appear to take a position different from that implied, but untheorized, by John Hunter.

In the fifth section of his lesson, however, Cuvier goes a step further than Hunter by linking the physiological and anatomical principles discussed above to natural history (taxonomy). Cuvier argues that while the naturalists have been correct in basing their taxonomic systems upon modifications of the major organs, where these are arranged in an order of importance as characters, they have paid insufficient attention to the functional significance of these modifications in making such assessments of importance. Here Cuvier hints at the second of his famous principles, the physiologically defined subordination of parts in an organism. He does not develop a definitive order of subordination in this text, but gives extended descriptions of the major taxonomic groups utilizing both structural and functional differences in a manner that recalls Hunter's recognition that the great classes display a uniformity of structure in the parts of greatest importance in their oeconomy. But whereas Hunter had limited his treatment of taxonomy to a critique of the systems of the naturalists, and had presented a number of "artificial" schemes whereby species could be classified according to a single index without acknowledging that any one of these could serve as the basis for a natural classification, Cuvier asserts that the hierarchical order of parts may be established on functional grounds. In the *Leçons* Cuvier takes various physiological indices, but in his lectures of 1807, and in a systematic manner in *Le règne animal distribué d'après son organisation* (1817), he developed a theory of the functional dominance of the *sensorium,* and divided the vertebrates into four *embranchements* according to the four general plans of the nervous system.[96] Thus, Cuvier utilizes arguments similar to those marshalled by John Hunter before him against the linear classificatory schemes of the natural historians, but goes further in fashioning comparative anatomy into the instrument of natural history in a manner that the Scottish anatomist explicitly denied was possible.

Disease and Death

I have discussed so far the principles of comparative anatomy that lay behind John Hunter's arrangement of the Physiological Series in his museum. The series in this division exhibited normal anatomical structures defined by the actions of the healthy animal. A considerable number of preparations in the rest of the museum were illustrative of morbid anatomy. These latter specimens testify to the tremendous amount of autopsy work conducted by Hunter. Although he did not publish a treatise specifically on the topic of pathological anatomy, Hunter devoted many pages of print to observations on the anatomical effects of disease. To the conceptual framework within which these observations were made, I now turn.[97]

John Hunter is generally recognized in the history of medicine for propounding a physiological theory of disease. Such a theory defines disease as malfunction, as a pathological modification of the normal (healthy) actions of organs and parts. Hunter indeed espoused such a theory, and in so doing made a radical break with the central concepts of earlier seventeenth- and eighteenth-century theoretical medicine.

According to the so-called "ontological" theory of disease, common to Hunter's medical predecessors and to many of his contemporaries, a disease is an entity visiting, or somehow in, the sick body. It is a species, whose presence may be inferred from the symptoms: from the nature and quantity of discharges from the body, from the flushing of the skin, and so on. The physician's task is to apply treatment proper to the particular disease entity as a whole, or refrain from intervention should the "history" of the disease require this; hence the initial problem is to ascertain which species of disease is present. Indeed, the very concept of disease as a species defined ontological theoretical medicine as nosology, the classification of diseases. The great nosologists of the period, from Thomas Sydenham (1624–1689) at the beginning of the classical age to Philippe Pinel (1745–1826) at its end, were concerned to theorize the method of deducing the natural history of a disease from the superficial symptoms apparent to the clinician on the body in which it was inserted. Symptoms were the language of disease by which the invisible pathological presence could be read.[98]

With Hunter the concept of disease is quite different. First, with the performance of autopsies, he brought the formerly invisible internal pathological lesions into the medical field of view. Second, although Hunter was still concerned with nosology, the symptoms by which diseases are classified are no longer signs of an entity, but of "dispositions" of particular diseased anatomical parts to "wrong action," to physiological malfunction.

The animal or part disposed to act, generally takes on action, which is of three kinds: natural, restorative when injured, and the diseased; each arising from the corresponding class of dispositions, and therefore divisible into the same. ... The ultimate and visible effect of disease is action; but this is not the disease, for the action is only an effect, a sign, a symptom of disease.[99]

Both normal (physiological) and deviant (pathological) actions come to perform the same epistemological function. They constitute the semiology of the properties of both vitality in general, and sickness in particular. There is no separate ontology of disease. Pathology comes to reside within the sciences of organic structure and of life, rather than within natural history, or the "botanical" ordering of species of disease.[100]

For Sydenham too, as for the early eighteenth century's major nosologist, François Boissier de Sauvages (1707-1767) of Montpellier, the etiology of disease is to be considered as a physiological cause-and-effect process (for the Hippocratic physician Sydenham, as humoral disharmony). Classification for these ontologists, however, cannot be based on causes, for causes are in principle unknowable. Sydenham and Sauvages were epistemological sceptics who held that symptoms, defining species of disease like characters of plants in botany, alone provide knowledge for nosology.[101] A very different approach to classification was developed in mid-century by the Edinburgh physician William Cullen, who, in a largely traditional symptomatic nosology, included a group of diseases classified by their causes. Classed under the heading *Neuroses,* these were considered to be caused by disorders in the "powers of sense and motion" of the nervous system.[102]

Following Cullen this functional basis for nosology was developed by two British physicians toward the end of the century. John Brown (1735-1788) of Edinburgh in his *Elementa Medicinae* (1780) classified all diseases in terms of his theory of the degree of "excitability" of the nervous system, while Erasmus Darwin (1731-1802) in the *Zoonomia* (1794-1796) described four general classes based on functions of the *sensorium*—Irritation, Sensation, Volition, and Association.[103] I shall speak more of these theories in the second part of this paper. At present one should note that the neural pathologies were highly speculative; neither Brown nor Darwin had any degree of commitment to investigate the supposed physiological processes of disease comparable to John Hunter's. While causes and functions became the basis for classification in these systems of medicine, and while John Brown in particular had nothing but contempt for the ontological nosologist's botanical description of symptoms, relying on measurements of the pulse alone, physiology cannot be said to have entered the theory and practice of medicine with neural pathology.

In contrast, those of John Hunter's texts which treat of pathology are detailed attempts to explicate what he termed "the physiology of disease." At the beginning of his first lecture on the principles of surgery, Hunter informed his students that his intention is "to begin with the physiology of the animal oeconomy in its natural or healthy state; and then to come to pathology, or the physiology of disease, which may be called the *perversion of the natural actions of the animal oeconomy.*"[104] He considers pathological actions to be deviations from normal to abnormal functioning in the diseased organ or part of the body. The student of surgery, therefore, must first gain an understanding of the physiology of the healthy body. It was not until he was over a quarter of the way through the course that Hunter commenced discussion of the general principles of disease, having given an account of organic matter and the "vital principle," a summary of the important organ systems, and discussed the principles of the blood and animal heat. His students had been forewarned:

Before I treat of the diseases of the animal body, which is the intention of these lectures, it will be necessary to give such general ideas of the subject [physiology], and to lay down such axioms and propositions, as will enable you to follow me through all the necessary descriptions of preternatural actions of the machine.[105]

The surgeon and physician must have a knowledge of the normal physiology of the animal body to practice their art. Indeed, so closely does Hunter connect disease with life, that he needs to caution the morbid anatomist in the interpretation of anatomical signs revealed by an autopsy. Since "diseases of an animal body are always connected with the living principle and are not in the least similar to changes which take place in the dead body," the pathologist must distinguish between the genuine effects of disease and the results of mortification and putrefaction, lest "all the advantage to be derived from opening dead bodies" be lost.[106] In order for the signs of previous disease, revealed in the inner organic structure of a cadaver, to be read correctly, the actions of life and death must be identified and distinguished.

Death became conceptualized and entered the field of positive knowledge in various forms with Hunter, in addition to that of morbid nosology. He conceives, for example, of death itself as a "stimulus," which causes the general stiffening of the muscles of a cadaver. He suggests that certain morbid actions may actually be caused by previous conditions in life:

Life is the preserver of the body from putrefaction, and when life is gone putrefaction would appear soon to begin. But this is not uniform; it is sooner in some cases than in others; therefore there must be some other cause than the simple deprivation of life to account for this difference of time. . . . there is an action in life which disposes the body for a species of putrefaction (or decomposition) when dead, and very probably death is the effect of this action in these cases.[107]

Death itself has ceased to be an absolute in the realm of metaphysics. It has become a natural process, and has entered the chronology of nosography as an object constituted by medical theory, divisible, and classifiable.

I shall divide violent deaths into three kinds: first, where a stop is put only to the action of life in the animal, but without any irreparable injury to a vital part, which action, if not restored in a certain time, will be irrecoverably lost. The length of that time is subject to considerable variation, depending on circumstances with which we are at present unacquainted. The second is, where an injury is done to a vital part, as by taking away blood till the powers of action are lost; or by a wound or pressure being made on the brain or spinal marrow while life remains in the solids sufficient for the preservation of the animal, if action could be restored to the vital parts. The third is, where absolute death instantly takes place in every part, as is often the case in strokes of lightning; in the common method of killing eels, by throwing them on some hard substance, in such manner as that the whole length of the animal shall receive the shock at the same instant; by a blow on the stomach; by violent affections of the mind; and by many diseases, in all which cases the muscles remain flexible.[108]

Just as life with John Hunter entered the science of animal organization to order the field of comparative anatomy, so death entered the science of disease.[109]

Monsters. One may now appreciate the conceptual links between John Hunter's theories of life, organization, and disease. But there is one more connection yet, in fact, an inversion. For, just as the physician must have a knowledge of the normal actions of the body, so too the physiologist may increase his understanding of the natural vital properties of parts by observing their unnatural actions in disease. A special instance of this is with teratology.

Hunter believed that monstrous deviations are law-like: "There must be some principle for those deviations from the regular course of Nature, in the oeconomy of such species as they occur in."[110] Each species has its own peculiar form; deviations from this uniformity—monsters—result from either "a defect in the first arrangement of the original matter," or from an inherent "susceptibility" to monstrous growth in the case of an accident.[111] In both accounts, monsters have a special importance in the science of life, for "some of their structures may explain something in the physiology of the more perfect animals."[112] Hunter argues that many parts of monsters explain nothing about the animal oeconomy in general. For example, "a supernumerary leg having vessels and nerves going to it, explains nothing in respect to either the use of vessels or nerves." However, he continues, "some of their structures may explain something in the physiology of the more perfect animals; just as the 'weight' in a clock might explain the use of the 'spring' in a watch, etc. . . ." Hunter here implies the evaluation of structural deviations on the most general functional grounds. "The only thing which they would tend to throw any light upon, is the principle of animal life." Significantly, Hunter makes this distinction between kinds of deviation to argue that only the functionally most important deviations need be considered in classifying monsters. Further, he suggests that monstrosities may explain the different states of life and sensation—the different "conditions of existence"—before and after birth. Monstrosities with malformations of the nervous system, for example, may survive *in utero*, but not after birth. To summarize the position sketched here, it would seem that for John Hunter, monstrous forms may illuminate the interdependent vital processes of the animal oeconomy, parts of which must be correlated anatomically to serve the normal purposes of life.[113]

In this respect, comparison with monsters, with other animal species, and with diseased bodies, all serve to explicate the nature of life and organization:

> Besides having recourse to many of the inferior orders of animals for the elucidation of some of the phenomena of the more complicated orders, we are also obliged to Disease for many of our hints on the animal oeconomy, or for explaining the actions of parts, for the wrong action of a part often points out what the natural action was, and itself gives an idea of life. Disease often corrects our imaginations and opinions, and shows us that such and such parts cannot have the uses commonly attributed to them, and therefore brings us a step towards the knowledge of the true use. Monstrosities can contribute to rectify our opinions in the same if not a more intelligible manner. A monster is either from a deficiency of parts, which can be produced from art (and often is from necessity, as in operations), or else from a modification caused by a wrong arrangement or construction of parts, which will produce an unnatural action, by which means the natural action may be known.[114]

Part II

I have outlined in the above section the empirical fields to which John Hunter made a significant contribution. At the level of practice his work as a comparative anatomist had no precedent, and remained unmatched in England for many years. At the level of theory, we discover a naturalist whose attempts to systematize his subject, to theorize the relationships between living forms and between anatomical parts, projected that field of knowledge toward the comparative science of the nineteenth century. As a pathologist and a surgeon, Hunter was instrumental in the new endeavor to apply physiological conceptions and theory to disease conditions, to treat diseases as malfunctions rather than as entities. Finally, considering Hunter's work in the life sciences as a whole, we discern that each area of inquiry— comparative anatomy, physiology and pathology—illuminates the others, that form, function and disease are mutually related concepts planted in the soil of a science that, less than a decade after Hunter's death, would be termed biology.[115]

However, my foregoing discussion has remained at a certain level of discursiveness. Most importantly it has left unspecified two basic concepts in Hunter's biological theory—the concepts of vital action and organization: the nature of life, and the living organism. I have discussed the role of functional considerations in the definition of anatomical series and the hierarchy of organ systems in the animal oeconomy, but have not considered "function" per se. In this section my aim is to uncover the conceptual base of John Hunter's theory of the living organism. I shall speak less of specific functions substantively than of the concepts of action and organism in their theoretical relation. I begin with Hunter's idea of life (his concept of the living principle) and conclude with his theory of the organism, the living animal in which vital actions are organized and integrated to produce the action of the whole. My focus moves therefore from life to organization. My ultimate purpose, however, is not to isolate these two concepts, but to demonstrate that they are both enclosed within a more fundamental concept—the concept of localized vital properties. With this concept life and organization are related, and the problematic of the whole field of Hunter's physiology established.

The Vital Principle

At first sight the problem I have posed for this section—explication of John Hunter's theory of the relation between life and organization—would seem to contradict a physiological principle that Hunter repeatedly emphasized, and for which he is best remembered: the principle that life does *not* depend upon organization. He believed that a living organism is endowed with a vital principle, and that this principle is a power *superadded* to the animal (or vegetable) machine. This vital principle confers the power of self-preservation upon the organism. Hunter adopts a typically vitalistic, tautological definition of life—since the living principle confers the power of self-preservation, life is the resistance to death.[116] The power of a living body to resist death is derived neither from the organic organization, nor from the material composition, nor yet from the action of the body. A body may retain, after death, both the organization and material composition of its former living state, but only in death will it putrefy. "Mere composition of matter does not give life; for the dead body has all the composition it ever had. Life is a property we do not understand; we can only see the necessary leading steps towards it."[117] The living principle may be said, in fact, to be superadded to the animal machine.

Amongst the evidence upon which Hunter based his conception of life as a superadded principle conferring the power of self-preservation were the results of some experiments upon eggs. Hunter refers to these experiments several times in his writings, and, as with all observations for which he claimed the rights of priority, is careful to specify when they were made: 1755–1756, early in his career, while studying the embryogenesis of the chick.

I then observed, that whenever an egg was hatched, the yolk (which is not diminished in the time of incubation) was always perfectly sweet to the very last; and that part of the albumen, which is not expended on the growth of the animal, some days before hatching, was also sweet, although both were kept in a heat of 103° in a hen's egg for three weeks, and in the duck's for four. I observed, however, that if an egg did not hatch, it became putrid in nearly the same time with any other dead animal matter; an egg, therefore, must have the power of self-preservation, or in other words, the simple principle of life. [118]

To further "prove a living principle," Hunter tested the effects of killing an egg by freezing, upon its power of resisting heat and cold. He concluded that a fresh (living) egg has the power of resisting heat, cold and putrefaction. Furthermore, an egg possesses this power "in a degree equal to many of the more imperfect animals...." [119] An egg has, in common with the lower animals, a minimal degree of organization, if not indeed, like some lower animals, a seemingly homogeneous composition; the organization of the dead egg does not apparently differ from that of the living; and finally, the egg is inactive.

Elsewhere, Hunter argues that since the animals of the lowest order are the hardest to kill (a polyp cut in half simply regenerates itself), the least organized are the most tenacious of life; but irrespective of organization, so-called "simple life" may vary in degree of strength between species. An eel, for example, is capable of surviving out of water for a considerably longer period of time than a mackerel, yet construction of the respiratory circulation, which provides the "perfect pulmonary blood" essential for the life of the higher animals, is the same in each fish. "It will follow," Hunter concludes, "that an animal which has a superabundance of this [vital] principle, will have, in the same proportion, less need of a perfect blood or even blood at all, and will retain the principle of simple life a much longer time than those who have less of it...." The "simple life" of an eel, then, must be stronger than that of the mackerel. [120]

As a further evidence of the nature of the vital principle, Hunter refers to the ability of living animal matter to resist the digestive power of the stomach. Improving on certain circus tricks perhaps, he suggests that,

If one could conceive a man to put his hand into the stomach of a lion and hold it there without hindering the digestive powers, the hand would not in the least be digested; and if the hand of a dead man was put in at the same time, whether separated or not from the body, that hand would be digested, while the other would not. [121]

His implication is that the power of resistance to digestion is conferred upon the living hand by the specific principle of life, a power superadded to organization. The stomach tissues themselves, by virtue of the same living principle, have the property of resisting their own destructive actions. It is in consequence of the absence of this principle in death, and not in consequence of diseased actions during life, that the stomach of a cadaver appears often-times unusually putrefied; the stomach of a cadaver, having lost its power of self-preservation, has commenced to digest itself.

John Hunter thus held the opinions that "organization and life are two different things"[122] and that life is "reducible to one simple property in every animal"—the power of self-preservation. The first property conferred upon animal matter by the superadded living principle is the power to resist death. The great historical importance of a vitalistic mode of explanation such as Hunter's is that it represents an attempt to specify the uniqueness of the living at the level of its very material, making a conceptual break with the iatrophysical reductionism of the previous century (with its baroque machinery of pulleys, levers, cords, and pipes). Abstracted from the rhetoric of a super-added principle, Hunter's concept of the power of an organism to preserve itself from putrefaction refers to a unique *property* of animal (or vegetable) matter: "Animal and vegetable substances differ from common matter in having a power superadded, totally different from any other known property of matter, out of which arise various new properties."[123] Hunter never tired of condemning those who conceptualized life in terms of inorganic matter and motion alone, of organization devoid of special vital power.

The actions and productions of actions, both in vegetable and animal bodies, have been hitherto considered so much under the prepossessions of chemical and mechanical philosophy, that physiologists have entirely lost sight of life; and perhaps they have been led to this mode of reasoning because these properties are much more familiar, more adapted to our understandings, and more demonstrable than the living properties of organized beings. But unless we consider life as the immediate cause of all actions occurring either in animals or vegetables, we can have no just conception of either vegetable or animal matter. No wonder, then, that the theories of the older physiologists are ill built and ill supported, their principles being false.[124]

Hunter would begin his surgery course with a lecture on the distinction between inanimate and animate matter, "in order to know when and where the animal principle acts, and when other principles [for example mechanical] are made use of in an animal body in accomplishing all the different purposes of life."[125] He defines three basic types of matter in the universe: common, vegetable and animal. The latter two arise out of the former, and animal out of vegetable according to a principle of the emergence of properties. Each type of matter is endowed with properties peculiar to it. Hunter regards the qualities of common matter—solidity, figure, and so on—as properties of that matter, and in keeping with contemporary natural philosophy, he includes among these properties certain active powers. The force of gravitation, for example, is due to an active power of attraction inherent in every particle of common matter. The nature of these active principles is unknown; they are properties inferred from the phenomena of bodies in motion, and are in essence inexplicable. [126]

Similarly, Hunter argued in his second surgical lecture, animal matter is endowed with its own unique properties: the principle of preservation, for example. Although he routinely described these particular properties of living bodies as derived from the union of living matter with the vital principle, a notion reminiscent of Stahlian animism, he also suggested on several occasions that they may arise from "just a peculiar arrangement of the most simple particles." [127] According to the latter view the principle of preservation, for instance, is like the magnetic property of a bar of iron, which does not arise from any apparent change in composition of the iron.

A bar of iron without magnetism may be considered like animal matter without life; set it upright and it acquires a new property, of attraction and repulsion, at its different ends. Now is this any substance added; or is it a certain change which takes place in the arrangement of the particles of iron giving it this property? [128]

Hunter in this passage wishes to promote the latter view, that the properties of animal matter, like the magnetic property of iron, arise as the properties peculiar to that combination or arrangement of the particles of which animal matter, or iron, is composed. Whichever *hypothesis* one accepts, however, spiritual or material, the nature of the concept of properties of animal matter is the same. The properties are in essence inexplicable, and the problematic of physiology remains, as I explain below, the description and explanation of vital phenomena in terms of the specific vital properties of the anatomical parts.

Thus when Hunter asserts that life does not depend upon organization, or even that life is a principle superadded to an organized body, he means that living matter is different from other kinds of matter, and that in its very nature as such lie the causes of vital phenomena. Beneath the rhetorical mode of expression lies a monistic and materialistic philosophy of life:

Matter being endowed with properties which become the cause of our sensations, and the modes of action of those properties being hardly known, these properties become the foundation of the idea of spirit, viz. a species of intelligent quality that presides over and directs the actions of matter. But, as causes and effects of matter seem to be entirely connected with matter itself, and to be a property inherent in and inseparable from it, and as these are becoming better known, the "presiding spirits" are every day vanishing, and their authority becoming less. Although "spirit" is a good deal exploded from having a share in the actions of common matter, yet it is still retained in animal matter; and, most probably, because the action of animal matter is much more extensive and has two states,—the living and the dead: and as there is no difference in the visible mechanism between the two states, it was natural to suppose that there was what is called an animating or living spirit. But matter can have some of its properties changed by very trifling circumstances. A piece of glass is transparent; but if that piece of glass be split, it will become less so; split it into three, still less so; and so on till it becomes the most opaque body that can be: and still the whole is composed of transparent glass: therefore, opacity in a whole does not give the least idea of the transparency of its parts.[129]

Hunter's conceptualization of the ontology of vital properties should on this evidence be situated within the broad context of mid-eighteenth-century vitalism. Speaking generally, there were two basic alternative theoretical positions developed in the eighteenth century in reaction to the preceding hegemony of iatromechanical physiology—vitalism and animism. It has become an historical commonplace now, as indeed it was commonly perceived then, that vitalism, as I have suggested, stands between the two dualist alternatives of mechanism and animism. For, whereas vitalism was, as Georges Canguilhem has succinctly stated, "la simple reconnaissance de l'originalité du fait vital," a recognition that vital powers are properties of living matter itself, animism retained the mechanist's material world of mere extension and motion, and animated the inert matter of the living body by the direct intervention of a transcendent soul in every vital action. This latter would be a hasty characterization of the critique of the sufficiency of mechanical explanations of life developed by the major theorist of animism, Georg Ernst Stahl (1659–1734) of Halle, in his *Theoria Medica Vera* (1708), to which vitalists such as Hunter were later to present a monistic alternative. Thus, summarizing this historical flux of doctrines and schools from the vantage of the first decade of the nineteenth century (the "Whig" interpretation is his own), Georges Cuvier praised Stahl's "ingénieux rival" at Halle, Friedrich Hoffmann (1660–1742), as the one who "commença, à-peu-près vers le même temps, à donner la première indication de la route intermédiare que l'on suit aujourd'hui, en cherchant à distinguer les facultés propres de chaque élément organique."[130]

Basic to Hunter's doctrine, of course, is a theory of the emergence of vital properties in the process of animal and vegetable nutrition. A physiological topic to which he often returns is the process of vegetalization in plants, and animalization in animals, whereby common matter, in taking on new properties, is transformed from one species of matter to the next.

Vegetables alone appear to have the power of immediately converting common matter into their own kind. Animals probably have not that power, therefore are removed further from common matter; so that a vegetable seems an intermediate step between common matter and animal matter.[131]

The physiological processes related to what we should term the general function of nutrition—the processes of production, distribution, and exchange that constitute the internal animal "oeconomy"—occupy a central role in Hunter's physiology. The sequence, which I describe in some detail below, begins, in the higher animals, in the organ of digestion. The first function of the stomach is to "animalize" common matter; its second is to confer upon the resulting newly formed animal matter specific vital properties, a process Hunter terms "vivification."[132] The blood—in fact the chyle prior to that— is thus considered to be alive; it is the result of the transformation of common matter into animal matter endowed with the inherent properties of life. Vegetables differ from animals in that any part of a plant is capable of "vegetalizing," or converting to vegetable matter, the fluids that permeate its structure; the plants have no specialized organ of digestion. The processes of digestion are themselves uniquely vital processes, being neither chemical, fermentative, nor simply mechanical in nature. Only when deprived of the living principle are animal or vegetable parts susceptible again to the "common" (chemical) processes of degradation.[133]

The Power of Action

In discussing John Hunter's conception of animal matter, I have considered a single property—the power of self-preservation. There is however another fundamental property that Hunter believes to be conferred upon some living animal matter by the vital principle—the power of action. Since Hunter as a physiologist and comparative anatomist was concerned, above all, to understand the living organism and its organization from the point of view of the action and uses of the parts, and given the conception of living matter outlined above, one would expect the concept of a vital power of action to play an important role in his biological theory. Indeed it does.

From the "arrangement" of particles of animal matter arises, most probably, the principle of preservation, although "Every individual particle of the animal matter . . . is possessed of life, and the least imaginable part which we can separate is as much alive as the whole."[134] The power of action, however, does not arise from this same "arrangement for preservation, which is life"—although this principle is its condition of possibility—but from a higher level of organization. It is in fact the power of anatomical parts, of a muscle fiber for example. In other words, just as the principle of preservation is the unique property of all living animal matter, the power of action is the unique property of a specific anatomical structure, which consists of "two or more living parts, so united as to allow of motion on each other." Hunter sketches a hierarchical organization of parts and properties:

A number of these simple acting parts, united, make a muscular fibre; when a number of these are put together they form a muscle, which joined with other kinds of animal matter, as tendon, ligament, composes what may be called an organ. Thus, too, by the arrangement of the living particles, the other organs of the body are formed, their various dispositions and actions depending on the nature of the arrangement, for action is not confined to muscle, the nerves too have action arising from the arrangement of their living particles. [135]

In fact each and every active part of the animal machine is endowed with its own peculiar power of action. Hunter argues that the principle of life has wrongly been compared to the spring of a watch, or the moving powers of machinery in general, which are only the cause of the first action.

[T]his is not the case with an animal; animal matter has a principle of action in every part, independent of the others, and whenever the action of one part (which is always the effect of the living principle) becomes the cause of an action in another, it is by stimulating the living principle of that other part, the action in the second part being as much the effect of the living principle of that part as the action of the first was of the living principle in it. The living principle, then, is the immediate cause of action in every part; it is therefore essential to every part, and as much the property of it as gravity is of every particle of matter composing the whole. [136]

This passage not only summarizes John Hunter's concept of the anatomically localized vital property and points our way towards an analysis of his concept of organism; it also displays his thought as representative of the vitalist mode in physiology, the dominant view since the middle of the century, in which the elements of analysis of organic action, or vital properties, were conceptualized as uniquely vital powers or forces. The following concepts structure this conception of life as a specific problematic of vital properties.

First, the vital property of active power renders each part a *causal* agent within the animal machine. Parts of the machine, nonetheless, as the word describing it suggests, constitute simply mechanical devices upon which the active parts operate their power. Thus, the powers are vital while their effects are mechanical—for example, the effect of the muscles upon the skeleton in producing progressive motion. Each organ, too, may be considered to be composed of both active vital parts and passive mechanical parts.

The heart may be considered as a truly mechanical engine; for although muscles are the powers in an animal, yet these powers are themselves often converted into a machine of which the heart is a strong instance; for from the disposition of its muscular fibres [active], tendons, ligaments, and valves [passive], it is adapted to mechanical purposes; and this construction makes it a complete organ or machine in itself.[137]

When Hunter uses the term "animal machine," he refers to the whole animal body, living and non-living, including all of its parts. The "animal machine," however, is wholly different from the mechanical devices commonly termed machines since it has motive agents, the causes of action, throughout.

Every machine has a power; a clock has either a spring or weight, and so on. In mechanics the parts are dependent on one another, so that some *one* power is necessary to put the whole in motion; in mechanics too there is commonly but one ultimate effect produced, whereas in an animal body there are a thousand. The powers therefore of an animal body are differently placed and circumstanced from what they are in an inanimate machine: it is not one power that is setting the whole to work, because if that were the case an animal's actions would always be the same, but he is at rest in one part, moving in another, and so on; and as this is the case, he must have power in every part, so that his powers are diffused through the whole animal, which is almost composed of powers.[138]

These vital, non-mechanical powers produce *mechanical* effects; many parts of the body have no power of action within themselves, but are moved by the active parts. One particular example of this combination, which Hunter discusses at length, is the relation between muscles and elastic ligaments. Elasticity is a property of common matter "having no power of action arising out of itself." In some parts, regions of the arterial system for instance, elastic ligaments act as "reciprocal elongators," opposing the contractions of the muscles.[139] The body is therefore a mechanical machine with respect to its action, but a living machine with respect to its powers. Hunter informed his students:

In treating of an animal body I shall always consider its operations, or the causes of all its effects, as arising from the principle of life, and lay it down as a rule that no chemical or mechanical property can become the *first cause* of any of the effects in the machine. But as all animals have form and motion, which motion is directed by that form, these motions become mechanical, so that every motion may be truly called mechanical. Mechanics are therefore introduced into the machine for many purposes. The living principle, however, in itself is not in the least mechanical, neither does it arise from nor is it in the least connected with, any mechanical principle.[140]

I have argued that for Hunter as for many of his contemporaries the physiological concept of vital property is equivalent to such concepts in chemistry and natural philosophy as the forces of affinity, attraction, or repulsion; this is nowhere more true than with the concept of the vital property of muscle as a causal power or *force*. Many eighteenth-century physiologists claimed a Newtonian rationale for this theory of properties, just as they referred to Newtonian sources for an inductivist methodology, and warrant for a phenomenalist epistemology. Albrecht von Haller (1708-1777), for example, whose experimental investigation of the anatomical localization of vital properties exerted a tremendous influence on later eighteenth-century physiological thought, believed that an inherent force in muscle fibers, a *vis insita,* may be deduced from the phenomena of muscle irritability, or response to stimuli:

What therefore should hinder us from granting Irritability to be a property of animal *gluten,* the same as we acknowledge attraction and gravity to be properties of matter in general, without being able to determine the cause of them. Experiments have taught us the existence of this property, and doubtless it is owing to a physical cause which depends upon the arrangement of the ultimate particles, though the experiments that we can make are too gross to investigate them. [141]

Second, the powers of action of living anatomical parts are genuinely non-mechanical, in that the motive response of the active part bears no mechanical relation, or proportionality, to the stimulus. A stimulus that causes a muscle fiber to contract, for example, simply activates the power of motion in the fiber itself. The immediate cause of the response is the vital property of the part, and not the stimulus. Thus, each active living part in the animal machine has a power of *self*-motion. "It is more in concord with the general principles of the animal machine, to suppose that none of its effects are produced from any mechanical principle whatever; and that every effect is produced from an action in the part; which action is produced by a stimulus on the part that acts. . . ." [142]

Third, just as it is sufficient to explain the phenomena of the organism's self-preservation in terms of an essentially inexplicable property of animal matter, it is sufficient to explain the phenomena of muscle action, for example, in terms of an inexplicable property inherent in the muscle fiber. The physiologist has no more knowledge of the nature of this property, of its ultimate cause, than the natural philosopher has of the active principles of matter—gravitation, repulsion, etc.—in terms of which he explains the phenomena of bodies in motion. Physiology is a science founded in the last analysis upon properties inferred solely from the phenomena of vital actions:

We see the body move. We go further; we see the parts that have within themselves motion, which is the immediate cause of motion of the whole; and we see how that motion can be excited. But all this does not give us the first cause of motion in those parts, nor does it explain the mode of action of the parts themselves. . . . A muscle is the power that acts, or has the power of action in itself.[143]

Fourth, while I have chosen muscle contraction as a model of vital action, the problematic of vital properties instigates the analysis of every action in the organism into the effects of localized vital properties. In fact, Hunter assumes that each part has its own specific property and the specific mode of action that property causes. In other words, physiological analysis is at the same time anatomical analysis. It is in this sense that the living body, as I mentioned earlier, is an *anatomia animata*. The active "elements" of the animal machine, so to speak, are the smallest anatomical parts that have an independent power and mode of action. With the concept of the vital properties of anatomical parts we have reached the point of articulation between physiology and anatomy, function and structure, life and organization. In the rest of this paper I aim to prove that it is this fourth aspect of John Hunter's concept of vital properties that reveals the level at which the whole field of his physiology is structured as a specific problematic. The concept of anatomically localized vital properties is determined by an analytic interpretation of organic relations, which itself transcends the mechanist-animist-vitalist debates over the nature of vital properties. I shall argue that the historical specificity of John Hunter's physiology, its character as a product of the dominant mode of eighteenth-century physiological discourse, is given not by what Hunter says of the nature of life, but in what he can and cannot say of the relations between life and organization, between function and structure.[144]

To interpret Hunter's doctrine of a superadded vital principle, therefore, one must appreciate that through it he expresses a theory of vital properties. The structure of this explanation of the phenomena of vitality is crucial to remember in interpreting statements such as the following from Hunter's last and most elaborate work (*Treatise*), wherein he endeavored to show that

organization and life do not depend in the least on each other; that organization may arise out of living parts, and produce action; but that life never can arise out of, or depend on, organization. An organ is a peculiar conformation of matter (let that matter be what it may), to answer some purpose, the operation of which is mechanical: but mere organization can do nothing even in mechanics; it must still have something corresponding to a living principle, namely, some power. I had long suspected that the principle of life was not wholly confined to animals, or animal substances endowed with visible organization and spontaneous motion: I conceived that the same principle existed in animal substances devoid of apparent organization and motion, where there existed simply the power of preservation.[145]

Susceptibility of Impression

Self-preservation is the power of living matter in general, but the power of action is a property of anatomical parts: each part has its own specific action as a result of its particular power of action. There is, in addition, a second property of each anatomical part, a property, in fact, upon which the action of the part depends. This is the susceptibility each active part has of impression, a "sensibility" to the particular stimulus that excites action in that part. Therefore, at the core of John Hunter's physiological theory are two concepts referring to the specific properties of active anatomical parts upon which all physiological processes are supposed to depend: "Life I believe to exist in every part of an animal body, and to render it susceptible of impressions which excite action."[146]

Both concepts are characteristic of physiological thought of Hunter's day. The power of action was generally termed *irritability* (or, taking muscle action as paradigmatic, *contractility*), the susceptibility of impression, *sensibility*. Hunter, however, objects to the term irritability; it suggests the concept "irritation," which he defines as the cause of unnatural or diseased action. A "stimulus" is defined as the "cause of increased natural action";[147] and Hunter uses the term "susceptibility of impression" rather than sensibility, reserving the latter to refer to conscious sensation. The two properties of living anatomical parts, irritability and sensibility, were however no less the basic objects of Hunter's physiological theory than they were, in various forms, of the eighteenth-century vitalist version of the problematic of vital properties. I shall examine Hunter's use of these concepts in more detail below.

"Susceptibility" is a term which carried special connotations in Hunter's usage. These are apparent in the term "sensibility" which expresses the same concept: namely, that animal parts are capable of sensing. The mouth of a gland, for example, was conceived as having a sense faculty akin to "tasting," allowing it to respond only to the appropriate fluids.[148] While Hunter used the concept of susceptibility with a seeming abandon, which frustrated his sober editors from a more positivist-minded generation, the philosophical context of mid-eighteenth-century physiology reveals to us its logic.

The dominant epistemology of Hunter's day was "sensualism," the theory that all knowledge is gained through the senses, that every idea in the mind results, in the last instance, from sensation. The mind, as the seventeenth-century physician and philosopher John Locke (1632-1704) put it, "comes naked into the world." No mental contents are innate. The dominant model of the human mind was a statue progressively animated as it is endowed with one sense after another, its psychic content gained through the increase of experience made possible by the increase of sensibility. The senses therefore enjoyed a privileged role in eighteenth-century epistemology. It is precisely this priority given to sensibility that characterizes the eighteenth-century physiological theory of vital action that Hunter's texts reveal. Each part has a sensibility to stimuli, and without such a sensibility the vital power of the part would not be stimulated. Thus, in physiological theory, the concept of sensation and sensibility functions analogously within the theories of both mental and bodily action. In fact, to give the concept its ontological reference, both the sensorial substance and animal matter were considered to be endowed with the power of sensation on account of their inherent property of sensibility. By the end of the century the dominant view could be captured in an impatient assertion such as the following:

Surely we have no need of proving once again that physical sensibility is the source of all the ideas and all the habits which constitute the moral existence of man: Locke, Bonnet, Condillac and Helvétius have demonstrated this truth to the fullest degree. Among those who are instructed and who make any use at all of their reason, there is no one who can raise the slightest doubt in this regard. On the other hand, the physiologists have proven that all the vital movements are the product of impressions received by the sensible parts: and these two fundamental results, brought together in a reflective examination, form but one and the same truth.[149]

In speaking of the vital properties of parts, reflecting common usage, Hunter surrounded the concept of sensibility with a mentalistic terminology. For example, he ascribes the action of the arteries in "disposing" of nutritive fluid to the tissues to their momentary "disposition or feelings": "When the blood is good and genuine, the sensations of the arteries, or the dispositions for sensation, are agreeable; then the offices of the arteries are carried on suitably to the intentions of nature."[150] In describing the wound-healing process, or union by first intention, Hunter argues that "whenever two parts which have an affinity—which are sensible of one another—come into contact, and the impression each receives is the same, the effect on both must be the same, and the desires from such sensations the same, like a kiss."[151]

We would be mistaken to treat Hunter's terminology as simply metaphorical; for the concept of sensibility (or "susceptibility of impression") refers to a property that may be predicated both of the matter of body and of mind. Indeed, Hunter believes that "daily experience shows us that the living principle in the body acts exactly upon the same principle with the brain. Every part of the body is susceptible of impression. . . ."[152] Hunter does not simply apply the same terminology to the actions of body and mind, but he suggests that the principle or mechanism, and also certain processes, are common to both. Of the latter, for example, he argues:

Memory, or recollection of past impressions, has, I believe, principally been applied to, or supposed to be an attribute of, the mind only; but we know that every part endued with life is susceptible of impressions, and also that they are capable of running into the same action without the immediate impressions being repeated. Habits arise from this principle of repetition of, or becoming accustomed to, any impression, and the same thing exactly takes place in the mind.

He continues, "Custom arises from external impressions, either in the mind or body...."[153] Nevertheless, Hunter is aware of the problems of nomenclature. "I have used the word *consciousness*," he apologizes, "because we have no language existing answerable to all my views on the animal oeconomy.... There are actions in the body which come the nearest to consciousness of the mind of anything that I can conceive, and therefore I make use of this word; but it is commonly applied by philosophers only to the mind."[154]

It remained for a later, more philosophically sophisticated group, the *Idéologues* of the Institut National in Paris at the end of the century, to take up the problem of classification and naming as an essential part of the physiological project initiated by the analytic interpretation of life and, specifically, by the problematic of vital properties.[155] In reading John Hunter, it is sufficient to remember that when he uses mentalistic terminology, when he speaks for example of an "instinctive provision in the parts," or asserts that "Every part of an animal that acts from impression, acts from an instinctive principle," he is referring to specific vital properties of particular anatomical parts (in these examples, "irritability").[156] While each individual anatomical part, each element of the animal compound, is endowed with its own unique properties which cause its own particular mode of action, every vital property may be classified under one of two heads. These are, as Hunter terms them, the "power of action" (irritability) and the "susceptibility of impression" (sensibility).[157]

Disposition to Wrong Action

I outlined in the first part of this essay how John Hunter's pathological theories are grounded in physiology. Hunter defined diseased actions as deviations from the normal actions of health, and pathology, in short, as the "physiology of disease." One would expect, in light of the foregoing discussion, to find this physiological pathology to be informed, at the conceptual level, by the same problematic of vital properties. Indeed this is so, and an examination of Hunter's concept of "disposition to disease" will shed more light on his concepts of "susceptibility of impression" and "power of action."

The vital principle, Hunter believes, confers upon an anatomical part the susceptibility of impression or "renders it susceptible of impressions which excite action." A living active part, in other words, possesses the inherent property of sensibility to stimuli. An inherent power of the part is stimulated to cause the particular action of the part. In his theory of disease, Hunter develops each of these concepts: stimulus, susceptibility, power of action, and action itself.

First, Hunter argues that there are two types of stimulus, as I mentioned above: stimuli, properly speaking, excite normal physiological responses; irritations excite pathological responses. Amongst irritants Hunter includes poisons, "morbific agents," and physical objects. Second, an irritant cannot act upon a part (or organ, or constitution) unconditionally:[158] the part must have a susceptibility to it. The sensibility of the part, to modify Hunter's language, must confer upon the part a susceptibility to that particular pathological irritant. Third, Hunter argues that the irritation disposes the part to a wrong or pathological action. The disposition thus created is, in a sense, a pathological modification of the power of action of the part—or as Hunter puts it, it is "a degree further than susceptibility"; it "must be sufficient to overcome the natural and habitual actions of the part or whole, just as force overcomes the *vis inertiae* of matter."[159] Finally, the disposition causes a diseased action of the part, an action different from normal, such as inflammation, or violently increased activity. In many cases the disposition for diseased action is caused by an unnaturally strong stimulation.

James Palmer, editor of Hunter's *Works*, objected to the concept of disposition. Not appreciating that Hunter's physiology is couched within the framework of vital properties, he could not accept Hunter's definition of disease as a latent disposition and not as the observable action.[160] For Hunter, however, pathological effects are *signs* of the invisible disposition, just as the actions of any part are signs of its inexplicable power of action.

The most simple idea I can form of an animal being capable of disease is, that every animal is endued with a power of action and a susceptibility of impression, which impression forms a disposition, which disposition may produce action, which action becomes the immediate sign of the disease; all of which will be according to the nature of the impression and of the part impressed.

That every action, whether natural, preternatural, or diseased, arises from this susceptibility of impression, I think we must allow, which susceptibility may exist without even disposition or action, if no impression or stimulus is made; for the disposition is only formed in consequence of some impression, and the action is only the consequence of the disposition being so strong as to incline a part for action rather than for resolution [that is, the salutary processes which return the part to health]. Therefore, action is the ultimate consequence of impression being made on a part endued with a susceptibility of impression and the power of action, which impression gives them the disposition to act.[161]

Nineteenth-century pathologists found themselves quizzical about another assertion, typically Hunterian in its boldness, yet one that illuminates the logic of vital action in Hunter for us. The very first subsection of Hunter's *Treatise* is entitled, "Of Diseased Actions, as Being Incompatible with One Another."[162] Here Hunter states: "As I reckon every operation in the body an action, whether universal or partial, it appears to me beyond a doubt that no two actions can take place in the same constitution, nor in the same part, at one and the same time; the operations of the body are similar in this respect to actions or motions in common matter."[163] Hunter goes on to argue that no two fevers can exist in the same constitution; nor can two diseases that have local effects, such as scrofula, scurvy, venereal disease, or smallpox, exist in the same part of the body at the same time. The reason for this impossibility of combined action—and for the nonexistence, according to Hunter, of such supposed "mixed diseases" as "pocky-scurvy," "rheumatic-gout," and so on—is that every part or constitution has its own specific vital power, and in disease, its particular disposition. That Hunter implies this even when he speaks of action rather than "disposition to action" is clear from his comment that "There are many local diseases which have dispositions totally different, but having very similar appearances, have been supposed by some to be one sort of disease, by others to be a different kind, and by others again a compound of two diseases."[164] The task of the pathologist is to infer the disposition of a part—its disease, as malfunction—from the actions of the part. Here, as an extension of the physiological analysis of vital properties, is the historical character of Hunter's physiological pathology.

Conversely, though first applied in pathology, Hunter's indisputable "axiom that two processes cannot go on at the same time in the same part of any substance" (a proposition concerning actions of every kind of matter), explains various phenomena in physiology, such as the fact that food in the active stomach does not decay spontaneously as it ordinarily would: digestion is "a process superior in power to that of fermentation."[165] One recalls that, for Hunter, in living animal matter the vital properties have replaced (through animalization and vivification) the original "common" or physical and chemical properties. He does not, therefore, speak of a tension between these different kinds of property in the same part, and the axiom discussed here implies that he would not. However, the vital properties of living matter are not considered by Hunter to be absolutely invariable, like the active powers of physics and chemistry. His concept of disposition to disease reveals this important aspect of the concept of vital property as I shall explain.

In discussing the susceptibility and disposition of parts to disease, Hunter makes explicit an important characteristic of their physiological equivalents (sensibility and irritability). Both are capable of modification in degree. In the pathological case, for example, the disposition to diseased action most often becomes, as a disease pursues its course, a disposition for restoration. In fact, the major part of Hunter's *Treatise* is devoted to the salutary processes—inflammations of various types—whereby pathological lesions are healed.

Disease is a disposition producing a wrong action, and it must continue this wrong action till the disposition is stopped or wears itself out. When this salutary effect, however, has once taken place, the state of the body becomes similar to that in a simple accident [for example, a gun-shot wound], viz. a consciousness of imperfection is excited which produces the action of restoration.[166]

The initial susceptibility of a part for disease is modifiable too, depending in complex ways on the momentary condition of the part, its "original nature" (hereditary disposition), the whole animal constitution including its mental state, and the external environment. Hunter makes use of the concepts of "habit" and "custom" to explain variations in susceptibility to disease. The inhabitants of cities or of prisons, for example, enjoy a decreased susceptibility to the noxious air of their environment, which would adversely affect new-comers. In general, Hunter suggests, inflammation ceases through "the parts adapting themselves in time to their present situation, which I call custom."[167] Similarly, a physiological response may become adapted to a stimulus through "custom," and muscles may increase in strength through "habit" (exercise). Repeating the standard doctrine of his time, John Hunter conceives of both normal and pathological variations in the vital forces.[168]

Finally, the pathological problematic of dispositions illuminates the importance of knowing diseased actions for the physiologist. "We are . . . obliged to Disease," Hunter asserts, "for many of our hints on the animal oeconomy, or for explaining the actions of parts, for the wrong action of a part often points out what the natural action was, and itself gives an idea of life."[169] The semiology of disease is continuous with the semiology of life, because the phenomena of both diseased and normal action signify a disposition, or property.

John Hunter himself provides a convenient summary of the sequence of reconstruction that I have undertaken:

I have endeavoured to show that an animal body is susceptible of impression producing action; that the action, in quantity, is in the compound ratio of the impression [stimulus], the susceptibility of the part [sensibility], and the powers of action of the part or whole [irritability] ; and in quality that it is according to the nature of the impressing power and the parts affected. I have also endeavoured to show that impressions are capable of producing or increasing natural actions, and are then called stimuli; but that they are likewise capable of producing too much action, as well as depraved, unnatural, or what are commonly called diseased actions. The first of these I have mentioned by the general term irritations; the depraved, etc. come in more properly in treating of peculiar or specific actions. [170]

The Internal Animal Oeconomy

Having discussed John Hunter's concept of vital action in general, it is time to consider his analysis of particular physiological functions. The substance of Hunter's physiological doctrine is devoted to a certain theory of the internal life of the animal, and to a specific concept of the organization of the internal organ systems. The problematic of vital properties will be seen to shape Hunter's theory of the animal oeconomy in determinate ways.

Simple Life. Hunter's theory of the organism predicates, first of all, a division of the animal machine into two parts. Although, as I have explained, Hunter refers to life as a single simple principle superadded to, or a universal property of, the living body, he refers more specifically to two different vital principles; or perhaps, more strictly, to two separate provenances of the vital principle. These he terms *simple life,* or the *living principle,* and *sensitive life,* or the *principle of sensation.* Each confers the power of preservation, and the susceptibility of impression, upon a different division of the animal machine. Simple life is the living principle of the body; sensitive life the principle of sensation of the brain and nerves, or "common sensorium."

This dualism, although traditional, has an important consequence for Hunter's concept of the organism. The dichotomy of principles provides a a formula whereby Hunter distinguishes voluntary from involuntary actions and sensible from insensible parts in the machine. Above all, it allows him to treat the actions of what he terms the "internal oeconomy" as virtually autonomous from the will, and nervous action in general. The internal oeconomy (organs of digestion, circulation, and respiration) and the "external parts" (organs of motion and reproduction) are also considered virtually independent systems. In fact, Hunter all but excludes the functions of the nervous system from his attention, and establishes the major project of physiology as analysis of what a later generation would term the "automatic functions" of circulation and nutrition. [171] It is for this reason that the blood and the organs of circulation, the stomach and the organs of digestion, occupy the foreground of so much of his physiological theory. My focus will therefore be on John Hunter's concept of the internal animal oeconomy.

Hunter makes a fourfold distinction among the actions of an animal. At the first level there are actions "common to every part alike, such as those which are employed in the internal operations of the machine, as growth, alteration, building up, taking down, etc., every individual part (the smallest conceivable) acting for itself only, which actions may be called immediate."[172] He places these actions at the ends of the vessels—the terminations of the arteries and the beginning of absorbents—where an exchange of materials takes place.

Second, there are actions of whole parts, or organs which are employed in supporting the first general function: "the stomach, lungs, heart, and other organs of life, may be said to be continually supplying materials with which the first are employed in building or repairing the system." These organs "may be called labourers, being subservient to the first, which, as being engaged in laying down and taking up parts, may be called the bricklayers." Hunter tells his surgery students that it is the first kind of parts, the "bricklayers," the ends of vessels, "which compose the movements of the true animal, being those which are immediately employed about itself. It is the operations of these which properly constitute the animal oeconomy respecting itself, and it will be principally these which I shall consider." Indeed, most of his lectures, as well as his major *Treatise,* deal with the subject of the interaction between the fluids and the solids of the body, the general functions, it would seem, of growth and nutrition (and the repair of wounds and lesions). A third class of actions of the animal includes those of the "more remote parts of the animal oeconomy," the actions of muscles in progressive motion for example.[173] Finally, there are the actions of the parts that do not have an inherent power of self-motion, the elastic ligaments, for example, or the valves in the veins.

Elsewhere Hunter repeats a twofold division of the animal oeconomy into the parts that respect its internal and external functions, propagation being included in the former, and digestion, because the stomach is acted upon by external objects, in the latter.[174] More commonly, however, he refers to the parts that receive their susceptibility of impression from the simple life as those involved in answering the immediate necessities of the continuation of existence, including digestion, circulation, respiration, and nutrition. Their actions proceed with a regularity and constancy that is independent of the will and insensible. They constitute a system independent of the nervous system; it is this that Hunter terms the internal animal oeconomy. "From extraneous stimuli [non-nervous, local] arise all our internal insensible actions, and these depend upon the principle of simple life. . . ."[175] Hunter is concerned to minimize the importance of the nerves and the will: "Much more has been given to the brain and nerves than they deserve. They have been thought to be the cause of every property in an animal body; that independent of them the whole body was a dead machine, and that it was only put in action by them."[176] On the contrary, each part and organ has an independent power of action, a specific "vital property" as I have termed it, by virtue of the simple life that confers upon the least conceivable part the property of life ("self-preservation"), and upon each anatomical part its faculty of sensibility ("susceptibility of impression"). "This principle of sensation is our director with respect to all other external actions, but has no absolute power over the internal oeconomy of the machine."[177] "We know," Hunter argues, for example, "of no power which the will or the mind has over the absorbents," or lymphatic system.[178]

Near the beginning of his first Croonian lecture on muscular motion Hunter notes that "The great question has hitherto been, whether a muscle is susceptible of impression without the medium of a nerve, or whether a nerve is in all cases necessary to its being called into action; for a stimulus must either affect the nerve which affects the muscle, or the muscular fibre itself must be susceptible of immediate impression from the stimulus."[179] He goes on to answer that a muscle would appear to be affected in both ways, and that "the modes of stimulating a muscle will be different according to the nature of the animal or of the muscle." There are four types of stimulus, three of which are independent of the nerves, being those internal or external stimuli that regulate the "internal machine," growth, breathing, hunger, and the "desire of propagating the species"; these three kinds of stimuli are common to both animals and plants. The fourth type, arising from actions of the nerves, and peculiar to animals, is twofold, "one in consequence of the nerves being impressed by external matter, the other from their being impressed by the brain, or sensorium." This leads Hunter into a discussion of the relations between voluntary and involuntary actions, but he fails to specify what the mechanism of involuntary nervous stimulation might be. Further, and typically, he makes no effort to relate these distinctions systematically. Nevertheless a picture emerges of a set of basic functions—the internal oeconomy—that may be influenced, via the nerves, by "passions of the mind," but that proceeds normally by the mechanisms of non-nervous sensibility, one part stimulating another either directly or sympathetically.

The notion of internal functions independent of the will and nervous action therefore adds a dimension to our picture of Hunter's concept of the vital power of action, and illuminates the logic and role of Hunter's concept of sensibility. For each internal part is endowed with a power of sensation, irrespective of whether that sensation is transmitted to the brain.

The internal susceptibilities, with the consequent impressions and dispositions, are, first, of want, and second, of repletion; and all the other internal operations of the machine arise naturally from these two, especially repletion, as digestion, circulation, respiration, secretion, and intercourse of the sexes, etc. But the first movement of these actions appears to require the impression of external matter, the powers of digestion being excited by food being thrown into the stomach, in consequence of which circulation, respiration, secretion, etc., all follow, arising out of the internal operations of the machine; all which have nothing to do with the sensitive principle, but are wholly dependent on the living principle.[180]

Oeconomy as the System of Internal Susceptibility. The question of the nature of sensibility of parts, of muscle fibers for example, had fired a major controversy in 1755, to which Hunter no doubt alludes in the passage above, between Albrecht von Haller and Robert Whytt (1714-1766) of Edinburgh. Whytt rejected Haller's contention that irritability is a physical *vis insita,* an inherent property of the glutinous material of muscle fibers, and maintained that muscular contraction depends upon the sensibility of the soul in the nerves; but he also rejected Haller's distinction between irritable and sensible parts, claiming that the sentient principle endows with sensibility many more parts throughout the body than Haller had recognized. John Hunter's theory of vital properties represents a third position, since like Whytt he subsumed irritability under sensibility, but like Haller he believed that an irritable response may occur without nervous mediation. His position is closer to that of the Montpellier vitalists, and also to that of Francis Glisson (1597-1677), who reintroduced the Galenic concept of irritability into physiology in the mid-seventeenth century. For Glisson, as for Hunter, the matter of each organic element or part is endowed with both an inherent *perceptio naturalis* (distinguished from nervous sensation) and a power he termed *irritabilitas.*[181] Likewise, according to John Hunter each part of the internal oeconomy has its own power to act and its own susceptibility of impression, irrespective of nervous action, and sensation of its own action in the brain.

There is in fact a structural correspondence, or formal analogy, between Hunter's theories of the internal oeconomy and of the *sensorium,* insofar as each constitutes a discrete system of particular parts and actions. The concept of sensibility provides a discursive basis for this analogy, as I shall now explain, but underlying it conceptually is an analytic interpretation of vital action. The analogy is threefold, respecting the "seat" or central organ of the living and sensitive principles, irritation as a causal factor in disease processes, and the material basis of sensibility.

I mentioned earlier that Hunter believed that the same material property accounts for the sensibility of both body and brain. In his last work, the *Treatise,* published in 1794, Hunter introduced a hypothesis that was coming into vogue at the time in physiological theory—the idea of a subtle substance diffused through the *sensorium* (brain and nervous system) providing a material substrate for the sensorial processes of sensation, volition, and so on, in what was otherwise considered to be a homogeneous unorganized structure.[182] Hunter himself suggests that there is a *materia vitae* diffused through *every* part of the animal which gives each the power that was previously attributed to the vital principle. He argues that, although the hypothesis "cannot be proved by experiment," this *materia vitae diffusa* explains the identical principles of sensibility of the body and the brain:

Every part of the body is susceptible of impression, and the *materia vitae* of every part is thrown into action, which, if continued to the brain, produces sensation; but it [the *materia vitae*] may only be such as to throw the part impressed into such actions as it is capable of, according to the kind of impression; so does the brain or mind.[183]

Hunter explains his concepts of "habit" and "custom" in terms of this hypothesis. Every active part of the animal machine is endowed with sensibility: "Thus the actions of life, of the nerves, of the mind, and of the will arise from impressions being made on each so as to affect their principles."[184]

The analogy between processes of the internal oeconomy and processes of the *sensorium* enables us to appreciate the logic of another concept referred to earlier that is related to sensibility—the concept of pathological irritation. One might argue, as I suggested in the first part of this essay, that Hunter's theory of pathogenesis, like the physiological intentions of his theory of disease in general, prefigures certain developments of the ensuing decades.[185] Having analyzed Hunter's problematic of disposition to disease, however, I suggest now that the theory is more akin to precisely those contemporary eighteenth-century notions I discussed above: the neural theory of disease of William Cullen (the close friend and correspondent of William Hunter in Edinburgh) which in the last years of Hunter's life was developed most prominently by John Brown, and after Hunter's death by Erasmus Darwin.[186]

In the influential Brunonian system of medicine disease is treated quantitatively as the degree of arousal or depression of activity in a part. The seat of the inherent vital property of irritability—John Brown termed it excitability—which confers upon a part the power of response to pathological irritation (or to the influence of the environment in general), is located (as in Darwin's system) in the sensorial material, which includes medullary nervous matter and muscular solid according to the doctrine of the *sensorium commune*.[187]

John Hunter contradicts this psychological theory of disease at the level of his physiological concepts of vital power and sensibility. These are properties not simply of the "nervous medullary matter," but of each active part of the body. Medicines act directly upon the parts, and not mediately through the nervous system (though in practice, one must admit, both systems tended to legitimize, through the concepts of arousal and depression, the traditional prescriptions of opium and alcohol).[188] No doubt it is at the exponents of neural pathology that Hunter directs his polemic: "the motions of the nerves have nothing to do with the oeconomy of the part; they are only the messengers of intelligence and orders."[189] Yet both systems were based upon the concepts of sensibility and the inherent power of vital response to irritation. Differences, therefore, between Hunter's concept of the susceptibility of a part to pathological irritation and the theories of disease etiology of Cullen, Brown, and Darwin, parallel the manner in which Hunter's concept of the physiological property of sensibility differs from Whytt's: sensibility both normal and diseased, for Hunter but for neither Whytt nor Cullen, is independent of nervous action.[190] John Brown's system of medicine, it should be noted, cannot strictly be subsumed within the conceptual structure I have termed the problematic of vital properties; at least, not insofar as Brown reduced all vital action to the effects of a single systemic rather than localized property, excitability. Further, insofar as Brown held life to be a forced state, a response to external agents, rather than the living organism's resistance to corruption, his vitalism diverged significantly from that of Hunter.[191] Erasmus Darwin's system, on the other hand, would seem to parallel John Hunter's in structural terms, in that both were founded upon the analytical determination of local perversions in susceptibility. In 1795 Thomas Beddoes, Darwin's friend and fellow-traveller, wrote the following praise of the *Zoonomia* in his notes:

His is a *symptomatologie raisonnée.* . . . He treats of symptoms, as they are called, each by itself; and so has few or none of those groups of symptoms, which, as they often occur together, have hitherto come to be considered as one state or being; in the same manner as any plant or animal is one single being consisting of parts. . . . Nothing but such an analysis can give any true idea of the nature of diseases. Each particular part has its various deviations from the healthy state, accompanied or not by deviations of other parts, just as individual susceptibility and associations determine. Such accidental combinations constitute many of the species of modern nosologists. . . . An analytical nosology is the alphabet of medicine. This alone can enable us to read the characters in which nature writes diseases. What mischief will ensue if synthetic nosology be regarded as anything beyond a very clumsy expedient for abbreviating language, is easily shown.[192]

My argument is that both neural pathology, at least with Cullen and Darwin, and Hunter's physiological pathology, though opposed systems in medical debate, presuppose the concept of vital and pathological action constituted by the problematic of vital properties.

The contemporary model of the brain in Hunter's time was that of an undifferentiated (internally homogeneous) organ of nervous material (the "medullary matter"). Mind was considered to be diffusely located in the brain, psychological transactions with organs of sense and the muscles being mediated through the sensorial substance extended to the parts in the nerves. Physiologists nevertheless assumed that this structure is the "seat of mind" or common center where sensations meet and from whence acts of volition emanate. Their language implied, as William Lawrence put it mockingly in 1818, "a central apartment for the superintendent of the human panoptican [*sic*]; or, in its imposing Latin name, a *sensorium commune.*"[193]

John Hunter subscribed routinely to this concept. Moreover, (and here I continue to pursue the analogy between the internal oeconomy and the nervous system), he asserted that within the internal oeconomy the "simple life" also has a "seat"–the stomach. By the processes of internal communication termed "sympathy," which I shall discuss in more detail shortly, the stomach is both sensitive to the condition of all the other parts in the internal oeconomy, and capable of impressing them according to its needs (or diseased condition).

[T]he stomach is the seat of simple life, and thereby the organ of universal sympathy of the *materia vitae* or the living principle. . . . We shall find it to be as much the seat of universal stimulus and irritability as the brain is of sensibility. . . . The stomach sympathizes with every part of an animal, and . . . every part sympathizes with the stomach.[194]

Just as the brain is the seat of the sensitive principle, the stomach is the seat of the living principle. In physiological terms this means that the actions of the whole animal oeconomy are conditional upon the function of nutrition. The organism must before all else supply its wants. In addition, the stomach seems extraordinarily sensitive to conditions in the body, diseases, and states of mind, too; witness the sickness, vomiting, or purging which follows a horrible story or the beholding of distressing circumstances. Conversely, parts may sympathize with the stomach. "A glass of cider in some," for example, "shall produce a flushing of the face immediately; and spirits, in the end, we know, produce inflammation and supporation over the whole face. . . ."[195]

In anatomical terms too the stomach "may be considered as the first part of the animal,"[196] "first" in the literal sense of an "order" since descending the scale of beings ever larger does the stomach become in proportion to the animal's size, until a simple polyp is reached, which consists of little more than a bag capable of propagation.

An animal can exist without any senses, brain or nervous system, without limbs, heart, circulation, in short, without anything but a stomach. . . . A polypus is a stomach and parts of generation in one; and the complicating an animal [*sic*] is no more than adding other parts for various purposes.[197]

In a second, genetic sense the stomach is the first organ of an animal oeconomy since the sympathies of the stomach preceded nervous communication between parts in the newborn child. In fact, sensation does not commence until after birth, "when a new oeconomy is set up, in which sensation is called in by the living principle for the support of the whole, beginning first in the stomach by sympathy, and then going on in a series of actions."[198]

Oeconomy as Circulation. I have discussed three features of the internal oeconomy as a system in which "the more simple actions arose independently of sensation or of the actions of the nerves."[199] Hunter's concept of the relation between the internal parts and the nervous system contrasts sharply with his notion of the relation between the parts and the blood. A denervated part or organ may continue to live and even function normally; but cut off its contact with the circulation, and it will surely immediately perish. The blood is the all-important part of the animal oeconomy. Upon its vitality depends absolutely the vitality of the whole organism. Having surveyed Hunter's general theory of the sphere and role of internal sensibility, we are enabled to consider what he regarded, to judge simply from the pages devoted to it, as the most important process in the animal oeconomy—the very economics of life. I shall now discuss the blood and its exchanges with the solid parts.

Hunter begins the first lengthy section of his *Treatise* ("General Principles of the Blood") with a consideration of the experimental evidence as to the cause of the coagulation of the blood. He concludes that physical factors in the arteries, the lack of motion of the blood, or its temperature, are not responsible for coagulation. In fact coagulation is an operation of life, a vital action dependent upon the living principle of the blood, and like any such operation—muscle contraction, for example—dependent also upon an impression. The stimulus arises in the solid parts according to their need for new solid matter. Coagulation, the process whereby living solids are produced from the blood, is for Hunter the basic essential general function of the organism. As Hunter argued before his students:

Coagulation is a species of attraction arising out of this irritation [impression]. It may be considered as a species of generation, for it is the first action or establishment of a power of action within itself, so as to form itself into muscular fibres, the only powers in an animal: these, again, with other parts, form organs to act on the materials from which they arose, for their own support.[200]

John Hunter believed that the blood itself is alive, that it is the "moving material of life" and that the motion of the blood is the "first moving power."[201] The same living principle or in later work the *materia vitae* that is diffused through the solids, is diffused through the blood. Blood is derived from chyle, which is the product of digestion and the first living fluid in the system. Vivification, or the endowing of animal matter with the principle properties of life, the powers of self-preservation and action, is, Hunter believes, second of two actions performed by the digestive organ upon food. Before the vital properties may be conferred upon food material, it must be converted into animal matter. This first process Hunter terms "animalization," and as a result of it extraneous material is given the unique "arrangement" of animal matter.

That the blood is fluid is no objection to its vitality, for organization is necessary only for gross actions such as contraction, not for life itself—the self-motive parts of the body are therefore the solids. As a homogeneous living fluid the blood is circulated in the arteries to every part of the body, and each part acts upon it according to its need. Where necessary a certain part of the blood, the "coagulable lymph," is extravasated from the arteries and coagulates to produce, first, a vascular tissue in which extravasation may continue, and then the particular solid part, which becomes endowed in the process with its specific vital properties as these arise from its organic arrangement. This is the sense in which Hunter argues that "blood is the material out of which the whole body is formed and out of which it is supported."[202] He has very little indeed to say about respiration, and has no concept of respiratory exchanges in the tissues. Of the pulmonary function of respiration he says simply that, upon the analogy of "animal fire," something in the air is necessary to ignite the vital principle latent in the chyliferous product of digestion carried to the lungs:

[T]his letting loose of the animal fire seems to depend upon the air, as much and in the same manner as common fire. So that instead of something vivifying being taken from the air, the air carries off that principle which encloses and retains this animal fire. I do not mean real and actual fire; but something that is similar, and is effected and brought about much in the same manner . . . so that the aliment we take in has in it, in a fixed state, the real life; and this does not become active until it has got into the lungs; for there it is freed from its prison.[203]

Living matter is thus transferred by extravasation and coagulation of lymph from the living fluid to the necessitous living solid part.[204] This, however, is but half of the interaction between solid and fluid. For the material of the solid parts is in a continuously dynamic state, being broken down and removed by that division of the vessels whose importance Hunter felt it was his great achievement to stress—the absorbents, or lymphatics. Hunter terms the action of the absorbents "modelling," because, in a growing part, "they are often taking down what the arteries had formerly built up."[205] By the combination of this process with growth, the form of the bones, for example, is modelled.[206] In disease whole organs, even, may be reduced and removed by the absorbents. Like all animal powers, absorption depends upon a stimulus acting upon the sensible part of the system, the mouths of the vessels.[207]

In the latter part of the eighteenth century, due largely to the work of William and John Hunter, William Hewson, William Cruickshank, and Alexander Monro *secundus,* the absorbents, including the lacteal and lymphatic vessels, came to be regarded as a complete system separate from but of equal importance to the vascular system. William Hunter offered the following definition in his lectures on anatomy:

I think I have proved, that the lymphatic vessels are the absorbing vessels, all over the body; that they are the same as the lacteals; and that these altogether, with the thoracic duct, constitute one great and general system, dispersed through the whole body for absorption; that this system *only* does absorb, and not the veins, that it serves to take up, and convey, whatever is to make, or to be mixed with the blood, from the skin, from the intestinal canal, and from all the internal cavities or surfaces whatever. This discovery gains credit daily, both at home and abroad. . . .[208]

William continued with no less an assertion than that this discovery is the greatest in anatomy and physiology since Harvey's discovery of the circulation of the blood, and he defended his proprietory claim to it through many bitter pages of polemic elsewhere; and for John Hunter, and the Hunterian school too, the doctrine of the lymphatics played a major role in theories of the animal oeconomy. John's experiments purporting to prove that the veins do not absorb remained the basis of this doctrine until it was disproved in the second decade of the nineteenth century, above all by the experimental work of François Magendie.[209]

With this summary view of the basic sequence of processes of the internal animal oeconomy, the sequence animalization-vivification-circulation-extra-vasation-coagulation-growth-modelling-absorption, one is in a position to appreciate the great importance of the vessels in Hunter's physiological theory. The vascular system is for Hunter, as one can see,

in some degree, to be considered as the efficient part of the whole animal respecting itself, every other part of the body being more or less subservient to it, and depending upon it for existence and support; and therefore the greatest attention should be paid to every circumstance that can possibly explain the various uses of the vessels, for there is no operation respecting the internal oeconomy of the animal but is performed by them, insomuch, that for the convenience of the vessels in performing those peculiar actions, they seem to constitute various combinations, which are called organs.[210]

Some organs, Hunter observes, in a manner that harks back to the fluidist physiology of Boerhaave, are so vascular that they appear to consist almost wholly of vessels.[211] He argues that generally the extravasated lymph coagulates to form vessels at first—an action of special importance in the several salutary processes that Hunter discusses in the ensuing chapters of his long *Treatise*. Take for example the most simple process of the healing of a lesion—inflammation and the union by first intention. An inflamed lesion is filled with extravasated fluid which coagulates gradually, drawing the two surfaces together. These surfaces are highly vascular, much lymph-like fluid being "exhaled." Union takes place by a uniting of opposed vessels, and these become the basis for continued circulation in the healed part. The absorbents, too, constitute a far more extensive system, Hunter claims, than had hitherto been realized. Indeed, from a general point of view the absorbents

may be considered as the animal consisting of so many mouths, everything else depending upon them or belonging to them, for in tracing these dependences we find there exists ultimately little else but absorbents. The stomach and the organs connected with it in such animals as have a stomach, are to be considered as subservient to this system.[212]

For John Hunter the internal oeconomy constitutes a general division of the animal body, a division that is virtually autonomous from the nervous system: the power of the "simple life" is independent of both nervous action and the will. Nevertheless the internal involuntary parts have nervous connections and may be stimulated by the sensitive principle. Further, the living principle sympathizes with the mind—witness the hunger instinct or general depression. "If it is asked why the involuntary parts have nerves at all, the answer may be given that it is not for their common actions, but to keep up the connexion between the whole, for without them an animal would become two distinct machines, and one might be acting very contradictorily to the other."[213] To unify organismic action, the living and sensitive principles must communicate with each other. In fact, "they become blended with each other, so that one seldom acts perfectly unless the other is in perfect order,"[214] harmonizing perfectly when the machine is in health. But one should remember that in the scale of nature and in the circle of life the living principle precedes the sensitive. Hunter suggests to the physiologist that

The involuntary contraction should be first considered, as the more necessary operations of the machine are carried on by it; for the machine could even exist independent of any voluntary contraction: but it could not go on if left wholly to the voluntary contraction of the muscles, unless we are endued with innate ideas capable of producing a will. This involuntary contraction is very extensive in the system, and is employed in carrying on a number of operations, of which the circulation is one; and which may be said to be, in a great measure, the oeconomy of the animal within itself.[215]

I have attempted to reconstruct John Hunter's theory of the "internal animal oeconomy" and have illustrated the nature of this organizing concept of the "oeconomy" above all as it designates the circulation of the subsistence materials of life. In general—and this point will be enlarged upon later—I have aimed to demonstrate that this theory is structured by the problematic of vital properties insofar as it ascribes every function through the mechanism of non-nervous sensibility to an organ or system, and each action, as an involuntary contraction, to a specific part. Descending through the various levels of organization of the oeconomy, there appear first subsystems such as the "stomach and the parts preparatory and subservient to digestion," and the vascular and lymphatic systems; next appear the organs; and finally the universal and necessary connections between tissues and vessels—"the animal consisting of so many mouths." Yet even here, at the level at which general physiology in the early decades of the nineteenth century would produce a new conceptual object—the general function of imbibition—Hunter can think only in terms of the specific action of parts. Indeed, he conceptualizes the nutritive process of material exchange with the aid of an anthropomorphism, ascribing the formation of fibers and membranes to the work of individual craftsmen—or rather to the work of both a bricky and his mate. Hunter's discourse cannot produce the conceptual objects of a general physiological theory of nutrition although its subject matter suggests at first that it should. Indeed, John Hunter's account of the internal animal oeconomy is perhaps not strictly part of a *theoretical* discourse at all, a set of concepts generated discursively in terms of general laws, but rather a classification of organic elements, a description of the stages of circulation between necessitous but independent parts. [216]

Many commentators have referred to Hunter's doctrine of the vitality of the blood, endeavoring to locate him in the vitalist tradition of William Harvey (1578-1657). All, however, have overlooked the fact that this "tradition" itself came into being within the chronological period and the theoretical context of the problematic of vital properties. In his treatise *De Generatione Animalium* (1651), William Harvey asserted that "both sense and motion are in the blood," and he considered the blood to be as much a "part" of the body as any other member or organ. [217] Like his contemporary Francis Glisson who introduced the concept of irritability into physiology, Harvey attributed the actions of parts to inherent faculties and powers. Harvey's doctrine of the vitality of the blood is an element of his monistic conception of animal matter; the spiritual principle of life is here conceptualized as having an immanent existence in the humoral fluid. [218] Similarly, John Hunter believed that the vital principle endows the blood with inherent powers of sensibility and irritability. "I have reason to believe that the blood has the power of action within itself, according to the stimulus of necessity." [219] Hunter's concept of the living blood, a fluid which "coagulates from an impression . . . influenced in a way in some degree similar to muscular action," is produced by, and functions within the problematic of vital properties.[220] It is this problematic that produces the objects of Hunter's experience, such as the active, sensitive blood as one connecting link, one agent, in the insensible, involuntary, gradual, and continuous process of circulation between the individual parts which compose the "internal oeconomy" of an animal. Thus, at the very beginning of the classical age, and at its end, there appeared comparable objects of discourse, fluidist modes of expression of the logic of vital properties.[221]

The Concept of the Organism

I have discussed in turn John Hunter's concepts of life as the principle of self-preservation; irritability, or the power of action of an anatomical part; sensibility, or the "susceptibility of impression"; and disposition to diseased action. Lastly, I have considered Hunter's concept of the "internal oeconomy," and have discussed the major processes of this system in some detail. My discussion has thus moved from the general principles of life, through the concepts of specific localized vital properties, to a theory of the actual organization of the internal animal functions (exemplifying the theoretical functions of those concepts); I have shown how Hunter conceptualized the internal life of an animal as the continuous, automatic circulatory processes of its "oeconomy." I propose now to stand back from Hunter's substantive theory of the animal machine and consider his general notion of intra-organismic relations.

Sympathy. According to the concept of the vital property that Hunter terms the "power of action," each living active part of the animal body has the power of self-motion inherently within it. Each part is therefore an independent cause of action in the animal machine, operating by mechanical effect upon the non-active parts. The corollary of this concept of anatomically localized vital properties is the concept of the animal machine as a compound of active elements, each acting according to its own peculiar power. This concept is applicable to two levels of organization; first, the level of the major organs, each answering a different purpose in the animal oeconomy; and second, the level of the vital elements that compose the organs, such as the fibers of a muscle, or the mouths of vessels, each possessing its own specific power of action.

An animal is, Hunter believes, "a compound of parts totally different in their sensations, stimuli, powers, and uses, from one another; each part doing one office, and no more."[222] An observation of Hunter on one of his pet subjects—the "oeconomy of the humble bee"—recalls a metaphor commonly used at that time to express this concept of the organism as a congeries of individual parts: "This insect is a striking instance of the union of the different parts of nature with each other, each part acting immediately for itself, yet collecting for others, and each depending on another, making in the whole one uniform machine, although made up of many and various parts." [223] Hunter's concept of the organism is in fact such an atomistic one, of a whole that is no more than the sum of its independent parts:

[A]n animal is composed of parts, each part, and each motion of each part, having its own moving power, capable of producing its immediate and remote effects independent of each other; so that in animals many effects may be going on at one and the same time, and each actuated by its own peculiar power, by which means an innumerable variety of effects are carried on at the same time, and without the least confusion or interference with each other. [224]

If "no two parts have the same action," how does Hunter conceive of their compounding in the whole animal, and the integration of their actions? In accordance with his concept of vital action, and the logic of his atomistic theory of the organism, the simplest mechanism he proposes is that of sequential stimulation.

To produce the ultimate effect in any machine, there must be a succession of actions, one naturally arising out of another, each part taking on the action peculiar to itself, the preceding action being always the stimulus to the next succeeding one; and thus the parts go on acting in regular succession until the ultimate effect is produced, and then the whole is at rest until stimulated into action again. [225]

One should remember that by the term stimulus Hunter means an impression of which the part is susceptible by virtue of its property of sensibility, and that the ensuing action is not a mechanical but an irritable response, caused by the part's own power of motion. The action of one part affects the power of action of another. Therefore, even at the level of this simple model of sequential stimulation, the compounding of actions is not merely a mechanical domino effect. In fact Hunter has at his disposal a concept to explain the more complex compounding of actions in the machine, of parts that need not be physically contiguous. This is the concept of "sympathy"; sensibility and sympathy in Hunter, as in much of late eighteenth-century physiology, are really two sides of the same coin. "By this principle of action, called sympathy, an action arises without an immediate impression in a secondary way. . . . This action without immediate impression is one of the most complicated principles in the animal body, especially the more complicated animals, because it is the compounding of actions." [226]

Hunter interprets the traditional doctrine of sympathy in terms of his concepts of sensibility and vital action. The action of a part may affect the power of action of another part even when the latter is not in contact with the former. Hunter seldom gives the nerves a role in communicating the sympathetic stimulus:

In the investigation of this subject [of sympathy], we shall find all the principles of action in an animal, even in the most complicated, have a connexion with one another; for instance, the living principle, the action of the nerves, and the mind: and that the same principle in one part shall be affected by the same principle in another; and this is the simplest kind of sympathy I can conceive. Thus, the living principle of one part sympathizes with the actions of life in another part, as must be the case in all animals which have nerves. [227]

The simplest kind of sympathy is still based upon the model of sequential action. It is the mode of coordination in plants, and in the simplest animals which have neither brains nor stomachs. "The succession of motions in a sensitive plant," Hunter announced in his lecture on sympathy, "are no more than a succession of sympathies, or a succession of stimuli in consequence of those preceding." [228] And if the sympathetic connections of the brain and stomach were eliminated in higher animals, the mode of coordination remaining would be that of vegetables, as above. [229]

The characteristic of simple sympathy is not that sympathizing parts are contiguous (they need not be—"contiguous sympathy" is a special case), but that their actions are similar. In the animal, a paradigm example of sympathetic coordination is compound action of the muscles, the result of similar but symmetrical antagonists. Hunter offers the striking illustration of a man balancing on a tight rope. He has, however, no theory of nervous reflex, rarely mentioning the nervous system in relation to any example of sympathetic action. In the case of simple sympathy he speaks merely of "a connexion of the living principle in the powers of one part with those of another, which may be termed a species of intelligence," independent of any mechanical construction. [230] Hunter makes use of the doctrine of sympathy mainly in his theories of disease, local and constitutional. However, "this connexion of every part of the body with the others, by sympathy, becomes one of the extensive principles of action in the animal oeconomy, indeed is the basis of most of the compound impressions, affections and actions." [231]

Differentiation of Properties. The foregoing discussion has referred to Hunter's concept of the organization of the higher animals. It is these, the vertebrates, that display the complex specialization of structures whereby each organ or part has a particular function in the animal oeconomy. Hunter, however, had a keen awareness of the scale of beings, and that many if not most organisms are very simply organized. "To one animal which has a brain and nervous system," he acknowledges, "you have ten thousand without them." [232] The polyp, for example, is simply "a stomach and parts of generation in one." It is fashioned into the form of a muscular bag, a stomach, to perform the most basic function, digestion. At the same time, however, this structure, a single organ of homogeneous animal matter, is capable of performing many of the "offices" of life which are performed by specialized organs in the higher animals: circulation, respiration, and so on. Descending the scale of beings, therefore, from man to the polyp, there is as we would put it (and Hunter did not use these terms) an increasing anatomical homogeneity and a decreasing physiological division of labor:

Nature has, in the most perfect animals, formed parts very distinct from one another, for all the different functions or operations of the body; whereas, in the more imperfect, she has huddled parts together, and made some serve two purposes, or has joined two into one. For instance, the ureters of the Bird enter the anus;—the penis, vasa deferentia or oviducts, enter the anus. Still more imperfect animals have heart and stomach in one.[233]

Hunter is always careful to remind his reader that discrete structures illustrative of a particular function may only be found in the more complicated animals.

Hunter's concept of the organism, and the coordination of its actions is, therefore, a product of the problematic of vital properties, which analyzes the physiological processes of the animal body into their component actions, and locates each action as that of an independent organ or part. The organism, according to John Hunter, is a compound of independently active elements; every function, in the last instance, is referred to an organ or part. At the same time, Hunter's concept of the organism is modified by his principles of comparative anatomy, specifically by the concept of progressive complexity in the chain of beings; but complexity increases in the scale of organisms by the anatomical differentiation of vital properties.

Analogously, Hunter believed that vital properties become differentiated from the general power of action of the whole organism in the process of individual growth and development. He adduced several examples as evidence. Transplantation experiments, of which Hunter performed several in addition to his regular practice of tooth transplantation as a surgical dentist, led to the conclusion that the "disposition in all living substances to unite when brought into contact with one another . . . is not so considerable in the more perfect or complex animals, such as quadrupeds, as it is in the more simple or imperfect, nor in old animals, as in young; for the living principle in young animals, and those of simple construction, is not so much confined to, or derived from, one part of the body."[234] Hunter adds that if a part of an older or more perfect animal is separated from the body, it dies sooner than is the case with a younger or simpler animal. From this it would appear that a greater degree of anatomical and functional differentiation entails a greater interdependency of parts, another example of Hunter's implicit concept of the physiological division of labor. (Here, as elsewhere, Hunter suggests that animals closest in kind to plants provide the model of organisms in which the living power "is capable of being contained equally by all the parts themselves," with no interdependency.)[235]

The phenomena of sympathy in disease also provide evidence. Hunter observes that children are much more prone than adults to suffer universal sympathy, or general convulsions, as the result of irritation. This is because the differentiation of vital properties is still incomplete in children; or, as he puts it, "local and partial sensation, and irritability, are not yet formed. . . ."

> But as the sensations and partial irritability begins to be formed, each part, in some degree acting for itself, acquires its own peculiarities; so that when a local disease takes place in a patient that is very young, it is capable of giving a general disposition to sympathize; but as the child advances, the power of sympathy becomes partial, there not being now in the constitution that universal consent of parts. . . . This arises from the different organs acquiring more and more their own independent sensations as the child grows older, and gradually losing the power of sympathizing with one another, so that by the age of six years few parts suffer but those immediately affected. . . .[236]

Such parallels between the differentiation of parts and properties in the series of comparative anatomy and in embryogenesis and development led Hunter to suggest elsewhere that they might be related by a general principle:

> If we were capable of following the progress of increase of the number of parts of the most perfect animal, as they first formed in succession, from the very first to its state of full perfection, we should probably be able to compare it with some one of the incomplete animals themselves, of every order of animals in the Creation, being at no stage different from some of the inferior orders [sic]. Or, in other words, if we were to take a series of animals, from the more imperfect to the perfect, we should probably find an imperfect animal, corresponding with some stage of the most perfect.[237]

Some historians have seen in this statement a prefiguration of the nineteenth-century theory of the recapitulation of phylogeny in ontogeny. My analysis, however, suggests that its meaning derives rather from the logic of an analytic than an organic–historical interpretation of life.[238]

Conclusion

John Hunter's theories of life, organization, and disease appear to be structured, then, at the level of conceptual analysis explored in this paper, by a determinate set of rules that define the historical character of his overall scientific project. In demonstrating the organization of concepts according to these rules in Hunter's anatomical and physiological problematics, I have developed an argument that must lead now to the conclusion that Hunter cannot be regarded as one of the "moderns" in biology. It should be apparent from the foregoing discussion that ultimately Hunter failed to cast his subject within the theoretical framework and through the conceptual categories of the discipline I defined in my introductory section as biology. What remains is the need to comment further on the nature of these rules, or "discursive conditions of possibility," which determine Hunter's discourse on living beings and define its historical specificity. First I shall repeat in general terms the evidence upon which my argument has been built.

In the first part of this essay I juxtaposed Hunter's critique of taxonomy with his theory of the animal oeconomy as the latter appeared in his reflections on comparative anatomy. I argued that implicit in Hunter's rejection of classification according to a single characteristic index was a theory of organic structure. Similarly I discussed how with Hunter an anatomical series is not simply a table of forms ordered according to their perceptible differences, but a set of structures defined according to the general physiological function performed. Thus, analogy and internal structure—and its physiological conditions—removed for Hunter the possibility of representing the order of things immediately in the order of words in classificatory schema. A theory of the animal oeconomy, absent from the taxonomic practices of eighteenth-century natural historians, arbitrates the arrangement of living forms in Hunter's comparative project. Yet Hunter did not pursue the taxonomic problem that was to preoccupy philosophical zoologists in the years following his death. He failed to develop the potential relations between his theory of a physiologically defined hierarchy of anatomical structures and a "natural method" of classification such as was developed by Georges Cuvier. Despite his implicit development of a biological conception of the organism, Hunter appears unable to transcend the medical problematic of organic analysis, the traditional comparative study of the structure and function of organs.

In the second part of this essay I endeavored to explain this conceptual limitation by examining in fine detail John Hunter's concepts of vital action and function, and his conceptualization of the relations between structure and function at the level of a physiological theory of the internal animal oeconomy. Here Hunter's analytic mode of thought becomes apparent. The biological unities of the major functions, which defined the relations of structure and function in anatomical series, are reduced to a quite different kind of conceptual object—the specific properties of individual organic elements. At this level Hunter's discourse appears to be structured by a specific problematic that poses for physiology the problem of analyzing vital actions into the effects of the discrete vital powers of anatomically localized parts—the properties of susceptibility of impression and the power of action. I argued that this problematic precluded the possibility of Hunter's developing a theory of the functional processes that constituted the nineteenth-century science of the animal economy—the general processes of metabolism, for example. It is in fact a "prebiological" problematic.[239]

The Problematic of Observation. There remains now the question of how this determinate structure of thought can be said, as I have claimed, to be an instance, embodied in the writings of John Hunter, of a general historical unity characterizing physiological discourse throughout the classical period. Michel Foucault has suggested that knowledge in this period, exemplified in the sciences of life, language, and the economy, took the form of the ordering of the complex representations of things according to systems of signs— *taxonomia*, as he has termed it.[240] The principal science of living forms was therefore natural history, which orders and classifies plants and animals according to their outward perceptible characteristics. Taxonomy, the *taxonomia* of living beings, involved the analysis of the complex appearance of plants and animals into perceptible characters, and their ordering and classification according to these immediate visible differences. The existence of this form of knowledge, according to Foucault, was due to the complete transparency of language to what it represented. Signification of the order of things in language was exact: the ordering of signs (language) was precisely the ordering of representations (phenomena), with no mediational problem of signification. Such a structure of knowing and speaking—which Foucault terms an *episteme,* the deepest level of knowledge uncovered by an historical "archeology"—may be regarded as a determinate epistemology of observation and analysis. This epistemology, William Albury has demonstrated, is constitutive of physiology also in the classical period. For physiology too, research into the nature of vital processes meant the analysis of phenomena into their least perceptible elements. Analysis, Albury argues, brings together the elements of structure and function in the concept of vital properties; the problematic of vital properties was thus a product of the classical epistemology of observation and analysis.[241]

The epistemological basis of the problematic of vital properties received its most concrete expression with Condillacian *Idéologues* such as Bichat in France at the end of the eighteenth century. Analysis and naming were for these methodologically self-conscious *savants* the fundamentals of every science: physics, chemistry, anatomy, physiology, pathology, or psychiatric nosology. Such methods, however, were the basis of the nonmathematizable science of physiology throughout the classical age, even where, as in the case of John Hunter, they did not explicitly instigate a deliberate program of research into the nature of vital properties. Hunter had no explicit experimental program of analysis as did, for example, Albrecht von Haller or Bichat. Yet one may observe in Hunter's work limits to experimentation produced by the classic problematic of observation.[242]

John Hunter has justly been regarded as a creative and committed experimentalist. There are few pages in his texts in which reference is not made to some experimentally produced observation, some purposeful pursuit of a phenomenon observed at first by chance. "In animals," he asserted, "we must observe the natural operations of the animal; and where opportunity does not serve us to observe Nature in her operations, we must put her in the way of [yielding the means of] observing those natural processes."[243] I have made reference, for example, to Hunter's experimental investigation of the routes of absorption, and in general terms to his ideological valorization of empiricism and opposition to rationalist systematizing. "I think your solution is just," he once wrote to his friend Edward Jenner, "but why think? Why not try the experiment?"[244] Nevertheless, Hunter makes several specific criticisms of certain kinds of experimental investigation, and these reveal his implicit epistemology of observation and analysis.

First, Hunter's problematic of observation does not accord a role for quantification or mathematics in physiology. He constantly raises physiological and methodological objections to iatromechanical speculation. "It appears to me impossible to ascertain the quantity of the blood in the body, and the knowledge of it would probably give very little assistance towards better understanding the oeconomy of the animal."[245] "The power of contraction of the ventricle must be within the strength of the artery; but it is hardly possible to ascertain what is the strength of an artery; nor, if we could, would it enable us to ascertain the strength of the ventricle. . . ."[246] "It is impossible to say what the quantity of blood is that is thrown out of the heart at each contraction."[247] Generally Hunter refers to the possibility of quantification in order to reject its efficacy or deny its practicality.

Second, the problematic of observation rejected the importance of chemical analysis—the only important differences between things were those that could be observed. Thus, nosology was based upon visible characters rather than chemical properties:

[E]xperiments have been made on pus from different kinds of sores, with an intention to ascertain the nature of the sore by the result of such analysis. That sores give very different kinds of pus is evident to the naked eye; and that the different parts of which the blood is composed will come away in different proportions, we can make no doubt. . . . We may also observe, that such kinds of pus change, after being secreted, much sooner than true pus. . . . From all this I should be apt to conceive that such experiments will throw little light on the specific nature of the disease, which is the thing wanted.[248]

What one must infer from the actions of inflamed or suppurative lesions,

Hunter argues here, but cannot deduce from chemical analyses of pus, is the "disposition" of the part to heal or to continue the malignant processes of a specific disease.[249]

Observation too had its limits. Like many of his anatomical contemporaries Hunter had a great distrust in the microscope. Only the immediate visual sense was to be trusted; only representations given immediately to the naked eye could be subject to analysis. Hunter criticizes microscopic examination of the blood globules:

I am led to believe that we may be deceived by the appearances viewed through a magnifying glass; for although objects large enough to be seen by the naked eye are the same when viewed through a magnifying glass which can magnify in a small degree, yet as the naked eye, when viewing an object rather too small for it, is not to be trusted, it is much less to be depended upon when viewing an object infinitely smaller brought to the same magnitude by a glass.[250]

Hunter's argument, that the microscopist necessarily lacks the cues of physiological adaptation of the eye to its object, reaffirms the epistemology of observation and analysis. "Concerning natural objects, we usually acquire a gross knowledge, from the frequency with which they are observed, and it often requires little more than common attention to have a tolerable conception of their general principles. This is the case with the blood."[251] Hunter's experimental and anatomical problematic, therefore, has little use for mathematics, chemistry, and the microscope:

[There have been] rather speculative philosophers than practical anatomists, [who] have frequently been misled with respect to the very facts and observations whose result was to decide the truth of their opinions. What, for instance, does it explain in digestion, that the force of the gizzard of a turkey is found equal to four hundred and seventy-three pounds? Does it afford a better solution to our doubts than we should derive from determining the force of the mill that grinds the wheat into flour? Or, on the other hand, will the most correct idea of fermentation allow us to account for the various phaenomena in the operation of digestion? But we can have no very high idea of experiments made by men who, for want of anatomical knowledge, have not been able to pursue their reasoning beyond the simple experiment itself.[252]

Correct reasoning, Hunter believes, infers from the phenomena of digestion and the cadaver on the dissecting table a power which is neither mechanical nor chemical, but the anatomically localized vital property of digestion.[253]

Finally, Hunter evidences a concern for the problems of nomenclature inevitable in a form of knowledge in which the order of representations is the order of nature:

In collecting animals, even the name given by the natives if possible should be known, for a name to a naturalist should mean nothing but that to which it is annexed, having no allusion to anything else, for when it has it divides the idea. This observation applies particularly to the animals which have come from New Holland; they are, upon the whole, like no other that we yet know of, but as they have parts in some respects similar to others, names will naturally be given to them expressive of those similarities, which has already taken place; for instance, one is called the kangaroo rat, but which should not be called either kangaroo or rat; I have therefore adopted such names as can only be appropriated to each particular animal, conveying no other idea. [254]

From the point of view of my analysis in the second part of this essay, then, in which Hunter's physiological problematic of vital properties is seen to be produced by the classical *episteme* of the eighteenth century, the concept of organic structure I elaborated in the first part should be considered as a product of the eighteenth century too, caught in transition as it were between the problematic of anatomical series (at one extreme *taxonomia*), and an explicit synthesis in the great laws of anatomy formulated by Georges Cuvier.

That Hunter's medical texts are still, despite the appearance in them of the positive concept of death, products of the classical age I have attempted to explain in my analysis of Hunter's concept of the disposition to disease. The same observation has been made of Xavier Bichat. "In the analyses of Bichat," Foucault argues, "the fundamental opposition of life and death was able to emerge." [255] Yet as Foucault admits, "Thanks to Bichat, super-ficiality now becomes embodied in the real surfaces of membranes. Tissual expanses form the perceptual correlative of the surface gaze that defined the clinic. By a realistic shift in which medical positivism was to find its origin, surface, hitherto a structure of the onlooker, had become a figure of the one observed." [256] William Albury has focused on this continuity of the epistemology of analysis in Bichat, and I have tried to show how his arguments have relevance for understanding John Hunter.

To conclude, therefore, I suggest that John Hunter's scientific work resides within a period of transformation in the most basic fabric of the knowledge of living beings. Aspects of his thought appear to prefigure the appearance of a science of biology in the first decades of the nineteenth century; but a recon-struction of the discourse in which that thought found expression reveals a mode of knowing and speaking of life fundamentally discontinuous with later developments. What appear to us at first as the evident conceptual possibilities of Hunter's comparative anatomy and physiology remain unrealized in his writings despite his major contributions to the enlargement of the empirical domains of these sciences, the internal anatomy and the internal animal oeconomy. By taking the structure of John Hunter's discourse as my ultimate object of analysis, I have uncovered the nature of the imperative behind this imposition of conceptual limits, this necessary silence.

Political Oeconomy. Finally, I would like to refer to my second statement of intent at the beginning of this essay. I suggested that the kind of method utilized in this reading of John Hunter is of value for the social history of science, in that it creates the possibility of situating scientific knowledge in its social context at a level deeper than that of the immediate social and institutional functions of specific doctrines, concepts, and practices. Put in other terms, my chosen method makes it possible to ask how the form and content of a *total* discourse—for example, the physiological discourse on organization and the animal oeconomy in the eighteenth century, and its underlying rules of construction—might be themselves, no less than the discursive elements of explicit debate, mediated expressions of human values and purposes, and products of historical and material practices. Viewing scientific discourse at this depth makes it likely, as Karl Figlio has suggested, that ideological conflicts mediated through the clash of scientific doctrines will be found to be held together, perhaps indeed made possible, by the ideological investment of concepts at a deeper level. For example, the familiar ideological characteristics of vitalism and materialism at the turn of the eighteenth century (political, social, and religious) are effaced, as Figlio has shown, at the level of the concept of organization common to both poles of doctrine. This concept does not, however, represent a template of value-free knowledge upon which values may accrete at higher discursive levels. It is itself an embodiment of value, a sign, simultaneously, of natural fact—anatomical structure—and human values—the evaluation of "degrees of animality," perfection, and the like. [257]

I would like now to pursue one strand of Figlio's argument, namely, that at the deeper levels of discourse concepts may flow between areas traditionally defined as discrete disciplines, both "scientific" and "non-scientific." Specifically, I suggest that the concept of an "oeconomy" in the eighteenth century is constituted within a general discourse upon organization in both natural science and social theory. It would seem that the same analytic problematic of organic relations underlies theories of both the living body and the body politic in this period. The following notes explore possible conceptual isomorphisms between animal and political oeconomy. [258]

A comparison may be made between conceptual elements of John Hunter's theory of the internal animal oeconomy and those of contemporary political oeconomy. First, the conceptual "individuals" that compose economic society for political oeconomists in the classical period correspond, as such, to the parts that compose the animal oeconomy of the physiologists. The former are economically self-motivated individuals, the latter anatomical parts with the power of self-motion. Second, integration of each individual's economic activity in the motion of the economy as a whole is achieved, according to political oeconomy, by the summation of individual self-interested transactions to produce a "natural harmony of interests." Similarly, John Hunter considers that the higher animals are "complicated, being composed of different parts, whose actions and effects appear complete in themselves, as the action of a kidney, a liver, etc; yet all combine to produce an ultimate effect in the machine, namely, the preservation and continuance of the species." [259] The concept of a "natural identity of interests" is opposed to the notion that the economy needs to be regulated by the state; we have seen that analogously Hunter minimizes the role of the *sensorium* and will in regulation of animal functions. Operation of the internal oeconomy, a continuous process of production, distribution and exchange, regulates itself by, as it were, an internal "invisible hand." If the stomach, as seat of the simple life, institutes certain higher order principles of sympathy, it does so from within the economic process, just as prices, wages, and rent, for example, mutually adjust ("sympathetically") according to the operations of the market.

The concern of eighteenth-century social theorists with the historical development of the division of labor parallels John Hunter's concern to specify which order of organism he is discussing. The philosophical history of the former illustrated the ever increasing social division of labor towards modernity, just as philosophical zoology for Hunter illustrates the increasing anatomical specialization of organs to perform specific physiological functions. In this sense the famous Comparative Method of the eighteenth-century historicists parallels the project of comparative anatomy.

The human body is what I mean chiefly to treat of; but I shall find it necessary to illustrate some of the propositions which I shall lay down from animals of an inferior order, in whom the principles may be more distinct and less blended with others, or where the parts are differently constructed, in order to show from many varieties of structure, and from many different considerations, what are the uses of the same parts in man. [260]

Finally, a major issue of debate in late eighteenth-century political oeconomy was the notion that an increasing division of labor in economic society would rigidify the latter, making individuals or groups within the system dangerously dependent upon others. This is a concern of the physiologically minded physician, too.

The more complicated a machine is, the more nice its operations are, and, of course, the greater dependence each part has upon the other; and, therefore, there is a more intimate connexion through the whole. *This holds good in society.* It also holds good in the animal oeconomy. The most perfect animals cannot be hurt in part without the whole suffering, while the more imperfect may be considerably hurt without the other parts suffering much. Thus we find that a man cannot lose a leg without the whole frame sympathizing with the injured parts, as if conscious of a loss; while a frog appears to be but little hurt. A snail, lobster, lizard, etc., can lose many parts which will be restored again. A polypus is still less hurt by amputation; for a new animal arises out of the wound or cut. So far we find a gradation from the animal to the vegetable. [261]

Such comparisons remain at a certain level of discursiveness. There may be a deeper level at which discourse on oeconomy presents a unified structure. Foucault's notion of a classical episteme suggests that the limits to what can be said may be isomorphic in animal and political oeconomy. At several stages in this essay I have shown that John Hunter's theory of the animal oeconomy is unable to constitute the theoretical objects of biology—organismic relations of function and structure. His problematic is conditioned by an essentially analytic and classificatory mode of organizing knowledge. Likewise, Keith Tribe, in a recent study of eighteenth-century political oeconomy founded on a concept of the history of science analogous to Foucault's, has demonstrated that the problematic of political oeconomy is discontinuous with that of nineteenth-century classical political economy. [262] The "oeconomy" in eighteenth-century texts is not theorized as an economy known to us as such through the categories of neoclassical theory. Agents in texts of political oeconomy are heads of households, or administrators of the state as a household—in other words, political rather than economic agents, individuals classified in the discourse of political oeconomy according to their individual power, rather than constituted, as in classical political economy, in terms of a theoretically defined economic function. The major preoccupation of political oeconomy was with the processes of circulation between political agents, much as John Hunter was concerned ultimately to specify the situation of each organic part (with a specific vital property) in the general process whereby nutrient is circulated in the internal oeconomy. On the other hand, Tribe argues, "economic discourse makes possible the constitution of specifically economic agents, agents that are not related to 'real persons,' rational calculators or whatever: they are constituted discursively by relations of capital and labour, categories which themselves are formulated discursively." [263] This description could be translated into the categories of biology, and describe equally well the formulation of the concept of an organic *economy* in the early nineteenth century. [264] There would seem to have been a profound transformation in the discursive possibilities of science in this period creating a tremendous proliferation, expansion and accumulation of organisms and bodies of all kinds.[265] The method of reading employed in this essay does not submit to an empiricist closure, and thus furnishes material for the subsequent historical explanation of an epochal mode of thought. For if eighteenth-century discourses upon oeconomy— natural and political—together provide evidence of such a hidden structure, then the kind of conceptual history undertaken in this paper may fairly challenge the sociological imagination in the history of science, and raise the question of the objects of discourse of that discipline itself as a problem.

NOTES

1. Professor Hedley Atkins, reply to the toast at the Hunterian Festival of the Royal College of Surgeons of England, 14 February 1967, *Annals of the Royal College of Surgeons of England* 40 (1967): 337. Operators everywhere may thus trace their roots; e.g., C. Frederick Kittle, "John Hunter, 1728-1793, the Architect of Modern Surgery," *Bulletin of the American College of Surgeons* 61 (1976): 7-11. Virtually any recent issue of the *Annals of the Royal College of Surgeons of England* will reveal the contemporary vitality of John Hunter.

2. Treated with caution the Hunterian Orations will, of course, with their accumulated historical wisdom and professional assessment, yield an extremely detailed commentary on those aspects of Hunter's achievements in surgery and biological science which receive no mention here. The orations are listed to 1959 in John Kobler, *The Reluctant Surgeon: A Biography of John Hunter* (Garden City, N.Y.: Dolphin Books, 1960), pp. 417-20.

3. A useful discussion of the Bachelardian roots of the epistemological category of the "problematic," and a critical assessment thereof, may be found in Stephen W. Gaukroger, "Bachelard and the Problem of Epistemological Analysis," *Studies in the History and Philosophy of Science* 7 (1976): 189-244, especially 202-6. Cf. Dominique Lecourt, *Marxism and Epistemology: Bachelard, Canguilhem, Foucault,* trans. Ben Brewster (London: New Left Books, 1975), passim.

4. Michel Foucault, *The Order of Things: An Archeology of the Human Sciences,* trans. anon. (New York: Vintage Books, 1973), pp. 226-32, 263-79. The essential pioneering study of biology and natural history at the turn of the eighteenth century is Henri Daudin, *Cuvier et Lamarck: les classes zoologiques et l'idée de série animale,* 2 vols. (Paris: Félix Alcan, 1926). Cf. the Foucauldian reading of biological history in François Jacob, *The Logic of Life: A History of Heredity* (New York: Vintage Books, 1976).

5. William Randall Albury, "Experiment and Explanation in the Physiology of Bichat and Magendie," *Studies in History of Biology* 1 (1977): 47-131. Cf. idem, "Physiological Explanation in Magendie's Manifesto of 1809," *Bulletin of the History of Medicine* 48 (1974): 90-99.

6. Karl M. Figlio, "The Metaphor of Organization: An Historiographical Perspective on the Bio-Medical Sciences of the Early Nineteenth Century," *History of Science* 14 (1976): 17-53.

7. Cf. Georges Canguilhem, "L'objet de l'histoire des sciences," in *Études d'histoire et de philosophie des sciences* (Paris: J. Vrin, 1968), pp. 9-23. I endorse Figlio's contention that explanations of how science "works" at the immediate institutional and

political level—as a social, cultural, or technological "function" of some sort—may fail to reach the deeper mediated materiality of natural knowledge; while at the other extreme, a structuralist approach to historical discourse may ultimately fail to escape from an idealist epistemology of science, as in Foucault's *The Order of Things.* For effective explorations of the institutional approach to the history of English physiology, see especially two studies that straddle the period under consideration in this paper: Theodore M. Brown, "From Mechanism to Vitalism in Eighteenth-Century English Physiology," *Journal of the History of Biology* 7 (1974): 179-216; Gerald M. Geison, "Social and Institutional Factors in the Stagnancy of English Physiology, 1840-1870," *Bulletin of the History of Medicine* 46 (1972): 30-58. For case studies in the social history of science utilizing versions of a functionalist sociological and anthropological approach, see Barry Barnes and Steven Shapin, eds., *Natural Order: Historical Studies of Scientific Culture* (Beverly Hills, Calif.: Sage Publications, 1979); of immediate relevance to this paper is Christopher Lawrence, "The Nervous System and Society in the Scottish Enlightenment," in ibid., pp. 19-40. Note that one of the editors of this collection terms his approach "bourgeois materialism" (p. 17), and would perhaps perceive the affinities of his work to Marxist historical materialism to be greater with the older functionalist ("vulgar") version than with contemporary approaches. On the nonidealist intentions of Foucault's "archeology" (amply realized in his recent work), see Michel Foucault, *The Archeology of Knowledge,* trans. A. M. Sheridan Smith (New York: Harper Colophon Books, 1976). Useful introductions to Foucault are Ian Hacking, "Michel Foucault's Immature Science," *Noûs* 13 (1979): 39-51; Devereaux Kennedy, "Michel Foucault: The Archeology and Sociology of Knowledge," *Theory and Society* 8 (1979): 269-90; Jean-Claude Guédon, "Michel Foucault: The Knowledge of Power and the Power of Knowledge," *Bulletin of the History of Medicine* 51 (1977): 245-77. A methodological discussion of "discourse history," in the Bachelardian tradition, may be found in a review essay that has considerable historiographic significance for this study of "animal oeconomy": Keith Tribe, "The 'Histories' of Economic Discourse," *Economy and Society* 6 (1977): 314-44. A concerted, ongoing effort to chart the path between the functionalism of both the mainstream sociology of science and vulgar marxism on one side and the idealist and scientistic tendencies of discourse history and structuralist marxism on the other may be found in *Radical Science Journal* (London, 1975-).

8. What appears in Foucault's *The Order of Things* as a quite sudden discontinuity—an abrupt epistemological break—has been theorized by him subsequently as a more gradual transformation, as a process: Michel Foucault, "La situation de Cuvier dans l'histoire de la biologie," *Revue d'histoire des sciences* 23 (1970): 63-69; and discussion, pp. 70-92, especially p. 86; trans. Felicity Edholm, *Critique of Anthropology* 13-14 (1979): 125-30.

9. For John Hunter's biography I have referred especially to the following: Everard Home, "A Short Account of the Life of the Author," in John Hunter, *A Treatise on Blood, Inflammation, and Gun-Shot Wounds* (London: George Nicol, 1794), pp. xiii-lxvii; Joseph Adams, *Memoirs of the Life and Doctrines of the Late John Hunter, Esq.* (London: J. Callow, 1818); Drewry Ottley, "The Life of John Hunter, F.R.S.," in *The Works of John Hunter, F.R.S.,* ed. James F. Palmer (London: Longman, Rees, Orme, Brown, Green, and Longman, 1835). The four-volume *Works of John Hunter* will hereafter be referred to as *Works.* Individual Hunter texts will be referred to by their particular title, but cited according to the pagination of *Works.* Stephen Paget, *John Hunter, Man of Science and Surgeon (1728-1793)* (New York: Longmans, Green, 1898); S. Roodhouse Gloyne, *John Hunter* (Baltimore: Williams and Wilkins, 1950); John Kobler, *The Reluctant Surgeon;* Jessie Dobson, *John Hunter* (Edinburgh: E. and S. Livingstone, 1969); idem, *Dictionary of Scientific Biography,* s.v. "John Hunter."

10. Hunter's texts are notorious for their awkward and obscure prose. Most opaque are those left in manuscript by Hunter and subsequently published in two volumes as *Essays and Observations on Natural History, Anatomy, Physiology, Psychology, and Geology,* ed. Richard Owen (London: John Van Voorst, 1861). These were left "pure and simple" by the editor, reproducing the transcript of Hunter's amanuensis, William Clift (1775–1849). Those of Hunter's works published contemporaneously are considerably more polished. Both of his first two books, on the teeth, are said to have been edited by the prolific "Dr. Syntax," William Combe (1741-1823). See Lloyd G. Stevenson, "The Elder Spence, William Combe and John Hunter," *Journal of the History of Medicine* 10 (1955): 182-96; H. W. Hamilton, "William Combe and John Hunter's Essay on the Teeth," *Journal of the History of Medicine* 14 (1959): 169-78. Of his third book, *A Treatise on the Venereal Disease* (London: No. 13, Castle-Street, Leicester Square, 1786; and *Works,* 2: 113–488), Hunter wrote to a friend that "in order to render the language intelligible, I meet a committee of three gentlemen, to whose correction every page is submitted." (Quoted by George E. Babington, *Works,* 2: 123.) The three were Drs. Gilbert Blane, George Fordyce, and David Pitcairn. Hunter's *Lectures on the Principles of Surgery* (*Works,* 1: 199-643) were transcribed by James Palmer, editor of the *Works,* from a student's set of shorthand lecture notes (see note 22 below), and compose the most accessible Hunterian text. The *Treatise on the Blood, Inflammation, and Gun-Shot Wounds* (1794; and *Works,* vol. 3) suffered editorially, perhaps, from being only part way through the presses at the time of Hunter's death. For bibliographic details see note 21 below.

11. Biographical details in the following paragraphs may be found in Samuel Foart Simmons, *An Account of the Life and Writings of the Late William Hunter, M.D., F.R.S., S.A.* (London: J. Johnson, 1783); George R. Mather, *Two Great Scotsmen: The Brothers William and John Hunter* (Glasgow: James Maclehose, 1893); R. Hingston Fox, *William Hunter, Anatomist, Physician, Obstetrician* (London: H. K. Lewis, 1901); George C. Peachey, *A Memoir of William and John Hunter* (New York: Henry Schuman, 1946); Fenwick Beekman, "William Hunter's Education at Glasgow, 1731-1736," *Bulletin of the History of Medicine* 15 (1944): 284-97; idem, "William Hunter's Early Medical Education," *Journal of the History of Medicine* 5 (1950): 72-84; 178-96; idem, "The Self-Education of the Young John Hunter," *Journal of the History of Medicine* 6 (1951): 500-15; idem, "Teacher and Pupil: The Brothers William and John Hunter from 1748-1760," *Bulletin of the History of Medicine* 28 (1954): 501-14; Charles F. Illingworth, *The Story of William Hunter* (Edinburgh: E. and S. Livingstone, 1967); Jessie Dobson, *Dictionary of Scientific Biography,* s.v. "William Hunter."

12. A convenient assessment of the relations between the major Scottish anatomists of the eighteenth century—Monro *primus* and *secundus,* Douglas, and the Hunters—may be found in the introductory essay of K. F. Russell, *British Anatomy, 1525-1800: A Bibliography* (Parkville, Victoria: Melbourne University Press, 1963), pp. 25-32. See also John D. Comrie, *History of Scottish Medicine* (London: Baillière, Tindall, and Cox, 1932), 1: 303-64; J. Glaister, *Dr. William Smellie and His Contemporaries: A Contribution to the History of Midwifery in the Eighteenth Century* (Glasgow: James Maclehose, 1894); K. Bryn Thomas, *James Douglas of the Pouch and his Pupil William Hunter* (Springfield, Ill.: Charles C. Thomas, 1964).

13. William Hunter, *Anatomia Uteri Gravidi Tabulis Illustrata. The Anatomy of the Human Gravid Uterus Exhibited in Figures* (Birmingham: John Baskerville, 1774). William first advertised the work in October, 1752. John's studies of the lymphatics and mode of descent of the testes were published in his brother's *Medical Commentaries, Part I* (London: A. Millar, 1762), pp. 42-49, 75-89; reprinted in *Works,* 4: 1-19; 299-307.

14. John Hunter, *Treatise on the Blood, Inflammation, and Gun-Shot Wounds,* Part IV, in *Works,* 3: 541–78.

15. London: J. Johnson, 1771; and *Works,* 2: 1–56. This physiological treatise was supplemented later with a clinical guide to surgical dentistry: John Hunter, *A Practical Treatise on the Diseases of the Teeth, Intended as a Supplement to the Natural History of those Parts* (London: J. Johnson, 1778; and *Works,* 2:57–111).

16. William Hunter figures in textbooks of medical history as a pioneer of educational modernism, through both the success of his private school, and his unrealized plan to establish a national school and museum of anatomy in England. E.g. Frederick F. Cartwright, *The Social History of Medicine* (London: Longman, 1977), pp. 49–50. See also William Hunter, *Two Introductory Lectures . . . To which are Added Some Papers Relating to Dr. Hunter's Intended Plan for Establishing a Museum in London* (London: J. Johnson, 1784), pp. 115–30; Stewart Craig Thomson, "The Great Windmill Street School," *Bulletin of the History of Medicine* 12 (1942): 377–91. No doubt William Hunter, an exponent of Scottish medicine and Parisian anatomy, and through him John Hunter were influential in the rise of the London medical schools at the expense of that of Edinburgh toward the end of the century. On William's method of anatomical instruction see Toby Gelfand, "The 'Paris Manner' of Dissection: Student Anatomical Dissection in Early Eighteenth-Century Paris," *Bulletin of the History of Medicine* 46 (1972):99–130. For descriptions of the institutionally powerful lineage of John Hunter's students (in the London teaching hospitals, and in Philadelphia and New York) see Ernest Finch, "The Influence of the Hunters on Medical Education," *Annals of the Royal College of Surgeons of England* 20 (1957): 205–48; T. H. Sellors, "Some Pupils of John Hunter," *Annals of the Royal College of Surgeons of England* 53 (1973): 139–52, 205–17; John Kobler, *The Reluctant Surgeon,* pp. 195–241; Betsy Copping Corner, *William Shippen, Jr.: Pioneer in American Medical Education* (Philadelphia: American Philosophical Society, 1951).

17. Much of the historical folklore on John Hunter concerns the numerous priority disputes that he fought with the characteristic zeal of that age of nascent professionalism and that undoubtedly reveal the importance of scientific proprietorship for professional success. A contemporary account of John Hunter as a surgical entrepreneur may be found in the remarkable anti-biography of a rival surgeon: Jessé Foot, *The Life of John Hunter* (London: T. Becket, 1794). On Hunter's professional life, see Jessie Dobson, "Some of John Hunter's Patients," *Annals of the Royal College of Surgeons of England* 42 (1968): 124–33; Lloyd G. Stevenson, "John Hunter, Surgeon-General, 1790–1793," *Journal of the History of Medicine* 19 (1964): 239–66. On the processes of professional formation in this period, cf. Toby Gelfand, "From Guild to Profession: The Surgeons of France in the Eighteenth Century," *Texas Reports on Biology and Medicine* 32 (1974): 121–34.

18. Arthur Porritt, "John Hunter: Distant Echoes," *Annals of the Royal College of Surgeons of England* 41 (1967): 1, citing R. L. Edgeworth. David Elliston Allen, *The Naturalist in Britain: A Social History* (London: Allen Lane, 1976), p. 45, claims, without citation, that Hunter was chairman of this club about 1765.

19. The first edition (1786) contained ten papers, the second (1792) twelve, revised and enlarged by Hunter. Richard Owen edited these for *Works,* vol. 4 (1835; preface by Richard Owen dated 1837), including the remaining ten papers from the *Philosophical Transactions of the Royal Society of London,* Hunter's unpublished Croonian Lectures, and two pieces published in *Transactions of a Society for the Improvement of Medical and Chirurgical Knowledge,* vol. 2 (London: J. Johnson, 1800), to which series Hunter made several additional contributions.

20. John Abernethy, *Physiological Lectures Exhibiting a General View of Mr. Hunter's Physiology, and of his Researches in Comparative Anatomy* (London: Longman, Hurst, Rees, Orme and Brown, 1822), p. 60; Max Neuberger, "British Medicine and the Göttingen Medical School in the Eighteenth Century, *Bulletin of the History of Medicine* 14 (1943): 457–463; G. Ten Doesschate, "Introduction," Petrus Camper, *Optical Dissertation on Vision,* facs. Latin ed. (Nieuwkoop: B. de Graaf, 1962), p. 15.

21. A chronological list of Hunter's published papers is to be found appended to Drewry Ottley "The Life of John Hunter," pp. 189–92; reproduced in Jessie Dobson, *John Hunter,* pp. vii–xi. A complete list of editions of Hunter's books, including the many translations, is W. R. LeFanu, *John Hunter: A List of his Books* (Cambridge: Cambridge University Press, 1946). A bibliography and selections from Hunter's writings appears in *Medical Classics,* comp. Emerson C. Kelly (Baltimore: Williams and Wilkins, 1940), 4: 401–527. The paper on the heat of animals and vegetables is reprinted in facsimile from the *Philosophical Transactions* in *Homeostasis: Origins of the Concept,* ed. L. L. Langley (Stroudsburg, Pa.: Dowden, Hutchinson, and Ross, 1973), pp. 29–71. A great many Hunterian manuscripts were put in possession of Everard Home following Hunter's death, but for reasons unknown Home chose to burn most of them in 1823. The majority of surviving manuscripts were published in the nineteenth century, above all in *Essays.* The best account of Home is Jane M. Oppenheimer, *New Aspects,* pp. 1–105.

22. James Palmer's edition of Hunter's *Lectures on the Principles of Surgery* (in *Works,* vol. 1) is a transcription of the shorthand notes taken by Nathaniel Rumsey in the lecture course of 1786–1787. This text is so characteristically Hunterian in style, and full of examples to illustrate points, that Palmer conjectures that the student had access to the Hunterian manuscript (Hunter always read his lectures). Palmer was able to compare these notes with several other sets at the Royal College of Surgeons, all less complete. The next most full set, taken in shorthand by James Parkinson in 1785, has also been published: *Hunterian Reminiscences,* ed. J. W. K. Parkinson (London: Sherwood, Gilbert, and Piper, 1833). John Hunter first gave lectures privately at his own house in 1773 (or possibly as early as 1770). In October 1775, he advertised a public course on the "Principles and practice of surgery." He gave this course every year until 1790, from when on it was read by his brother-in-law Everard Home. See J. M. D. Olmstead, "A Student's Notes on John Hunter's Course of Lectures of 1775," *Bulletin of the History of Medicine* 7 (1939): 996–1003; "Syllabus of a Course of Lectures on the Principles of Surgery, Delivered by Mr. John Hunter, F.R.S. St. James's London," *The European Magazine* (October 1782), pp. 246–47; reprinted as an appendix in John Abernethy, *Physiological Lectures,* pp. 341–47.

23. Lester S. King, ed., *A History of Medicine* (Harmondsworth: Penguin Books, 1971), p. 27. For evidence that Hunter cultivated his un-bookish image deliberately, see "Anecdotes of Mr. John Hunter," *The European Magazine* (October 1782), p. 248: "He totally rejected books, and took up the volume of the animal body." Reprinted in John Abernethy, *Physiological Lectures,* pp. 347–52. Abernethy states (p. 201) that Hunter himself composed these anecdotes. A brief commentary upon the contents of Hunter's private library, revealed in more than 540 books sold in auction at Christie's in February 1794, is to be found in W. J. Dempster, "Hunter the Scholar," *World Medicine* 10 (1975): 87–96.

24. *Lectures,* pp. 207–8.

25. The second volume of Hunter's *Essays* consists almost entirely of anatomical descriptions of a large number of animal species. Hunter makes frequent reference in these to his sources of the particular specimens. One encounters: "a large black monkey"

from Lord Shelbourne; a genette "sent from the Cape of Good Hope to Mr. Aiton of Kew, who gave it to me," "a small ruminating animal, from the Prince of Wale's Island called a deer: given me by the Duke of Portland"; a siren from Dr. Garden of Charles Town (*sic*), South Carolina; and so on. Clearly Hunter's science benefited from the social contacts established through his surgical practice. Hunter obtained specimens from the menagerie at the Tower, and from various animal dealers at considerable expense. A useful connection was his friend John Ellis, famed naturalist and London agent for the Irish Linen Board with commercial interests in Dominica and West Florida. See Roy A. Rauschenberg, "John Ellis, F.R.S.: Eighteenth Century Naturalist and Royal Agent to West Florida," *Notes and Records of the Royal Society of London* 32 (1978): 149–64; Jessie Dobson, "John Hunter's Animals," *Journal of the History of Medicine* 17 (1962): 479–86.

26. Everard Home, "A Short Account of the Life of the Author," p. xxxi. When John White returned to England from New South Wales with several preserved specimens of the strange "nondescript" creatures of the southern continent, "There was no person to whom these could be given with so much propriety as Mr. Hunter, he perhaps, being most capable of examining their structure, and making out their place in the scale of animals. . . ." Dr. [George] Shaw, introduction to John Hunter, "Description of Some Animals from New South Wales," in John White, *Journal of a Voyage to New South Wales* (London, 1790), "Zoological Appendix"; quoted in *Observations,* p. 481. Hunter's skill at dissection was frequently employed by others; his dissections of *Gymnotus Electricus* and the torpedo, for example, were commissioned by the circle of Royal Society "electricians." W. Cameron Walker, "Animal Electricity before Galvani," *Annals of Science* 2 (1937): 92. Hunter also performed autopsies for his medical friends, and was a recognized authority on the embalming of bodies. He was summoned on several occasions to testify as an expert witness in court. See "Mr. Hunter's Evidence on the Trial of John Donellan, Esq.," *Works,* 1: 194–98. T. R. Forbes, "Two New John Hunter Manuscripts," *Guildhall Studies in London History* 1 (1973): 24–27. On Earl's Court see Cecil Wakeley, "John Hunter and Experimental Surgery," *Annals of the Royal College of Surgeons of England* 16 (1955): 69–93.

27. See Stephen Paget, *John Hunter,* pp. 153–61; John Kobler, *The Reluctant Surgeon,* pp. 274–80. Hunter printed his own editions of the *Treatise on the Venereal Disease,* of *Observations,* and of *Treatise,* probably to circumvent the common practice of printers who pirated cheaper editions in Dublin. On this and the Lyceum Medicum, see A. H. T. Robb-Smith, "John Hunter's Private Press," *Journal of the History of Medicine* 25 (1970): 262–69. Anatomists in eighteenth-century London were frequently the objects of popular abuse, occasionally of mob violence. The complex factors that formed the proletarian Londoners' image of the anatomists in the first part of the century, from the former's resistance to the legal right of the company of Barber-Surgeons and the Royal College of Physicians to the hanged bodies of Tyburn convicts—and the efforts of hospital and private surgeons to bribe such bodies into their possession—to folk-beliefs about death and its rituals, are excellently described in Peter Linebaugh, "The Tyburn Riot Against the Surgeons," in Douglas Hay et al., *Albion's Fatal Tree: Crime and Society in Eighteenth-Century England,* (New York: Pantheon Books, 1975), pp. 65–117. In the second half of the eighteenth century it was the anatomy schools' practice of obtaining their essential supply of cadavers from grave-robbers that directly offended popular sensibility. Perhaps, then, it was from fear of assault that John Hunter equipped the entrance to the building that housed his dissection room with a drawbridge. In considering the social relations of anatomy in this period one should, I believe, temper the point of view of the surgeons with, for example, William Blake's caricature of "Jack Tearguts" (in *Songs from an Island in the Moon*).

28. See Richard D. Altick, *The Shows of London* (Cambridge: Harvard University Press, 1978), pp. 27-28, 229. The museum did indeed leave Hunter's widow with tremendous debts, and his executors Matthew Baillie and Home endeavored to dispose of it to the British Government, unsuccessfully because the Treasury was currently funding a war. Finally, in June 1799, the House of Commons voted a sum of £15,000 to purchase the preparations, still kept at Castle Street and tended by William Clift, Hunter's last and devoted assistant. In December the museum was taken into the custody of the Company of Surgeons (which received a charter from the Crown the following year) and William Clift was appointed Conservator. In 1806 the museum was moved temporarily to a house adjoining the Royal College of Surgeons in Lincolns Inn Fields, finally occupying the collegiate building in 1813; that same year Baillie and Home endowed the annual Hunterian Orations. A condition of the Royal College's trusteeship of the museum was that a course of lectures in comparative anatomy be given each year. Richard Owen, who became one in the influential sequence of such lecturers, was appointed Assistant Conservator in 1827 to aid Clift in the overdue task of completing the catalogues of the physiological series of preparations. The museum remained the scientific basis of the surgeons' Royal College until a bomb crashed through the roof in 1941, destroying over half the collection (by then considerably augmented). In 1963, however, a restored Hunterian Museum was opened to the public, arranged according to the plan of the original catalogues. See Drewry Ottley, "The Life of John Hunter," pp. 137-56; Jessie Dobson, "The Hunterian Museum" in Zachary Cope, *The History of the Royal College of Surgeons of England* (Springfield, Ill.: Charles C. Thomas, 1959), pp. 274-306; idem, *Conservators of the Hunterian Museum* (London: Royal College of Surgeons, n.d.); G. Grey Turner, *The Hunterian Museum: Yesterday and Tomorrow* (London: Cassell, 1945); Victor Negus, *History of the Trustees of the Hunterian Collection* (Edinburgh: E. and S. Livingstone, 1966).

29. F. Wood Jones, "John Hunter's Unwritten Book," *The Lancet* 261 (1951), 2: 778-80.

30. Frederick Ruysch, *Thesaurus Anatomicus,* 10 vols. (Amsterdam, 1701-1715). The most remarkable illustrations from this series are reproduced, with commentary, in Antonie Luyendijk-Elshout, "Death Enlightened. A Study of Frederick Ruysch," *Journal of the American Medical Association* 212 (6 April 1970): 121-26. The auction catalogue of Richard Mead's collection reveals the full range of refined taste in a doctor-connoisseur—rare books, paintings, ancient coins, "antiquities"—while *opera naturae* compose a list of barely five pages: *Museum Meadianum, sive Catalogus Nummorum, Veteris Aevi Monumentorum, ac Gemmarum, cum Aliis Quibusdam Artis Recentioris et Naturae Operibus* (Londini: A. Langford et S. Baker, 1755). William Hunter shared many of these scholarly enthusiasms as a collector (as did John to a degree), and the extent of his patronage of auction sales, and art and rare-book dealers, is amply revealed in his museum, which was removed to the University of Glasgow in 1807. Even in that half of the space devoted to natural history, the display of whole animals—natural curiosities from around the globe—dominates the anatomical preparations. See J. Laskey, *A General Account of the Hunterian Museum, Glasgow* (Glasgow: John Smith, 1813). Nevertheless, William Hunter's anatomical collection was as famous in its time as was John's subsequently, as may be judged from Vicq-d'Azyr's florid praise of it in his *éloge* on the elder Hunter: *Oeuvres de Vicq-d'Azyr,* ed. J. L. Moreau de la Sarthe (Paris: L. Duprat-Duverger, 1805), 2: 382-83. On William's practice as an anatomist see John H. Teacher, ed. *Catalogue of the Anatomical and Pathological Preparations of Dr. William Hunter. In the Hunterian Museum, University of Glasgow* (Glasgow: James Maclehose, 1900), 1: xxi-lxxvii, "Introduction." Most sections of this catalogue list a subsection of comparative preparations. William's goals, however, remained traditional in anatomy;

he informed his students that "The human Anatomy is what we propose to explain: the comparative will only be introduced occasionally, where it serves to illustrate the other, or to guide us in reasoning from analogy." William Hunter, *Two Introductory Lectures,* p. 3. Further descriptions of the Hunterian Museum, Glasgow, are to be found in George R. Mather, *Two Great Scotsmen,* pp. 88–118; and R. Hingston-Fox, *William Hunter,* pp. 35–41. William's manuscript catalogue of his preparations was published in the nineteenth century as *Catalogue of the Anatomical Preparations in the Hunterian Museum, University of Glasgow,* ed. G. Fordyce, D. Pitcairn, W. Combe (Glasgow, 1840). For general discussions of anatomical museums in the eighteenth century, see F. J. Cole, *A History of Comparative Anatomy* (London: Macmillan, 1944), pp. 433–63; K. F. Russell, *British Anatomy,* 0p. 32–35.

31. I have referred to the following descriptions and catalogues of the museum: Everard Home, "A Short Account of the Life of the Author," pp. xxxviii–xlv; idem, *Lectures on Comparative Anatomy; in which are Explained the Preparations in the Hunterian Collection* (London: G. and W. Nicol, 1814), 1: 1–23; John Abernethy, *Physiological Lectures;* Drewry Ottley, "The Life of John Hunter," pp. 156–88; *Synopsis of the Arrangement of the Preparations in the Gallery of the Museum of the Royal College of Surgeons. For the Use of Visitors* (London: Carpenter, 1818); *Catalogue of the Hunterian Collection in the Museum of the Royal College of Surgeons,* 6 parts (London: R. Taylor, 1830–1831); *Descriptive and Illustrated Catalogue of the Physiological Series of Comparative Anatomy Contained in the Museum of the Royal College of Surgeons in London,* comp. Richard Owen, 5 vols. (London: R. & J. E. Taylor, 1833–1840); *Descriptive Catalogue of the Pathological Specimens Contained in the Museum of the Royal College of Surgeons of England,* comp. E. Stanley and J. Paget, 5 vols. (London: W. Clowes, 1846–1849); *Descriptive Catalogue of the Pathological Series in the Hunterian Museum of the Royal College of Surgeons of England,* 2 vols. (Edinburgh: E. and S. Livingstone, 1966–1972); *Descriptive Catalogue of the Physiological Series in the Hunterian Museum of the Royal College of Surgeons of England,* 2 vols. (Edinburgh E. and S. Livingstone, 1970–1971); *Hunterian Museum,* a guide prepared by Elizabeth Allen (London: Royal College of Surgeons of England, 1974).

32. Georges Cuvier, *Historie des sciences naturelles pendant la deuxième moitié du XVIIIe siècle et une partie du XIXe* (Paris: Fortin, Masson, 1843), 1: 191–94. Cf. idem, *Rapport historique sur les progrès des sciences naturelles depuis 1789* (Paris: L'Imprimerie impériale, 1810), pp. 320–21. Cuvier's lists include G. L. Leclerc Buffon (1707–1788) and P. S. Pallas (1741–1811); important also was the work of J. F. Blumenbach of Göttingen. See Paul L. Faber, "Buffon and Daubenton: Divergent Traditions Within the *Histoire naturelle,*" *Isis* 66 (1975): 63–74. John Hunter stands in the same relation to Alexander Monro *secundus* in comparative anatomy as he does to his brother William. While the practice of the two latter represented the peak of human anatomy in Edinburgh and London respectively, John Hunter's work evidences an interest that was separating itself from medical concerns, and becoming more nearly zoological. Monro designed his study of fish, for example, as a contribution to the essentially medical debate over the lymphatics, hoping that it would be "not less acceptable to the Physician than to the Naturalist." Alexander Monro, *The Structure and Physiology of Fishes Explained, and Compared with those of Man and Other Animals* (Edinburgh: Charles Elliot, 1785), p. 13. On the seventeenth-century comparative anatomists see F. J. Cole, *A History of Comparative Anatomy,* passim; M. F. Ashley Montague, *Edward Tyson, M.D., F.R.S. 1650-1708 and the Rise of Human and Comparative Anatomy in England,* Memoirs of the American Philosophical Society, vol. 20 (Philadelphia: American Philosophical Society, 1943).

33. Georges Cuvier, *Leçons d'anatomie comparée,* 5 vols. (Paris: Baudouin, 1800–1805). On the development of a "rational anatomy," see William Coleman, *Georges Cuvier, Zoologist* (Cambridge: Harvard University Press, 1964), pp. 26–106; Georgette Legée, "Les lois de l'organisation d'Aristote à Geoffroy Saint-Hilaire," *Histoire et nature* 1 (1973): 3–25. My interpretation in this paragraph follows Foucault. *The Order of Things,* pp. 263–79.

34. Richard Owen, "Preface" (1837), *Observations,* p. xl. Cf. ibid., p. vi: "Had Hunter published *seriatim* his notes of the structures of the animals which he dissected, these contributions to comparative anatomy would not only have vied with the labours of Daubenton as recorded in the *Histoire Naturelle* of Buffon, or with the Comparative Dissections of Vicq d'Azyr which are inserted in the early volumes of the *Encyclopédie Méthodique* and in the *Mémoires de l'Académie Royale de France,* but they would have exceeded them both together." If Owen was the foremost expert on Hunter of his day, that other British follower of Cuvier, William Lawrence (1783–1867), could probably lay claim to the broadest scholarly knowledge of recent biological history. Consequently, Lawrence's favorable comparison of Hunter's work with Cuvier's is still worthy of note: William Lawrence, *The Hunterian Oration* (London: John Churchill, 1834).

35. I have taken the respective numbers of preparations cited in the following paragraphs from Drewry Ottley, "The Life of John Hunter," pp. 158–86; Jessie Dobson, "The Hunterian Museum," pp. 304–5; idem, "Lost Treasures," *Annals of the Royal College of Surgeons of England* 42 (1968): 387–93.

36. Hunter's teratological theories, and the special conceptual role of the deviant animal form in his theories of life and organization, are discussed below. Within the discipline of his science, however, Hunter shared in the enthusiasm of his age for natural curiosities; indeed, in the context of his insatiable desire for new and unusual specimens to dissect and display, his interest in oddities merely adds a piquancy–witness the oft-told story of how he obtained the skeleton of an eight-foot Irish giant–to the general picture of his comparative anatomy.

37. Owen attributed this collection to the backlog of specimens, "accumulating as the reputation of the illustrious Founder increased and extended, and as the requisite leisure for their examination became abridged"; and he correctly disavows the importance of differences in outward form for Hunter, a crucial point in assessing the distance of Hunter's project from the dominant methods of the naturalists. See *Catalogue of the Contents of the Museum of the Royal College of Surgeons in London,* Part 4 (London: R. Taylor, 1830), "Advertisement," p. iii.

38. John Hunter submitted two papers to the Royal Society on fossils, of which only the first was published, and prepared a manuscript introduction to the catalogue of his fossil collection. John Hunter, "Observations on the Fossil Bones Presented to the Royal Society by His Most Serene Highness the Margrave of Anspach, etc.," *Observations,* pp. 470–79; Richard Owen, "Lectures Explanatory of Hunter's Manuscript Essay 'On Extraneous Fossils,' and Introductory to the Hunterian Course 'On fossil remains,' Delivered in the Theatre of the Royal College of Surgeons of England, in 1855," *Essays,* 1: 281–340; John Hunter, *Observations and Reflections on Geology. Intended to serve as an Introduction to the Catalogue of his Collection of Extraneous Fossils* (London: Taylor and Francis, 1859). See also *Essays,* 2: 500–502; *Descriptive Catalogue of the Fossil Organic Remains Contained in the Museum of the Royal College of Surgeons of England,* 4 vols. (London: Taylor and Francis, 1845–1856).

39. *Descriptive . . . Catalogue of the Physiological Series* (1833–1840).

40. Ibid., 1: ix, emphasis added.

41. Ibid., vols. 1–4.

42. *Synopsis . . . for the Use of Visitors* (1818), p. 2.

43. *Descriptive . . . Catalogue of the Physiological Series,* 1: 187–226.

44. *Observations,* p. 127.

45. Ibid., pp. 126–30.

46. Ibid., p. 130.

47. "Account of the Organ of Hearing in Fishes," ibid., pp. 292–98.

48. Ibid., p. 293.

49. Hunter does not suggest that he was thus led to investigate the organ of hearing in fish; indeed, he claims to have discovered this organ in 1760, and relates as evidence that fish can hear his observation of the startle response to gun-fire of fish in a pond in Portugal in 1763. My aim is simply to illustrate the potential consequences of the concept of anatomical series, to stress its a priori epistemological status.

50. *Treatise,* p. 101.

51. Background details in the famous portrait of John Hunter by Sir Joshua Reynolds have long puzzled Hunter scholars, but may illustrate this distinction. On the table beside Hunter is portrayed a book, open to display two facing plates. The first shows a series of forelimbs, from the horse to the human hand; the second a series of skulls from the human European, through the human aborigine, the chimpanzee, the macaque monkey, the dog, and finally to the crocodile. The series of skulls perhaps signifies the order of whole animals in the chain of being (see note 64 below). See Arthur Keith, "A Discourse on the Portraits and Personality of John Hunter," *British Medical Journal* 1 (1928): 205–9. Not surprisingly this detail has generally been "read" recurrently, and hence meaninglessly from an historical point of view, through the categories of evolutionary theory. On the other hand, John Abernethy's assertion may well be correct, that the illustrations are from Reynold's portfolio (presumably added after the sitting), and demonstrate facial angles and Reynold's interest in the then highly fashionable science of physiognomy, Hunter's opinions on which being nowhere, to my knowledge, recorded. John Abernethy, *Physiological Lectures,* p. 85.

52. Quotations in this paragraph are to be found at *Essays,* 1: 9–10.

53. Ibid., 1: 37.

54. Ibid., 2: 1.

55. Ibid., 1: 131.

56. Less strictly Hunter will speak of a general "Progression and declension of perfection in animals," wherein degrees of perfection are signified by the "complication" of animals considered as the sum of all their irregularly progressive parts. Ibid., 1: 36–37.

57. Ibid., 1:25. Hunter offers the following definition of a species: "Animals breeding in the full extent of that process constitute the species, although they may differ in some of their parts or other circumstances; but which [differences] are less essential, only constituting a variety." Ibid., 1:12. He collected information about cross-breeding dogs and wolves, and mated jackals with dogs at Earl's Court to determine that by the above definition wolf, jackal, and domestic dog are "three great varieties of the same species": *Observations,* pp. 319–30. Hunter was concerned here to determine the "original species," of which the wolf, jackal, and dog are historical varieties. He argues that the point of reference must be the "character of the genus . . . for it is natural to suppose that the original animal, or that which is nearest to it, will have more the true character of the genus, and a stronger resemblance to the species nearest allied to it, than any of the other varieties of its own species" (p. 327). Since the wolf has the greatest resemblance to the fox, it should be regarded as the original; domestication of the wolf has led to the dog, and jackals are probably dogs returned to the wild. Animals in the wild, especially in cold climates, show little variety; geological and climatic changes,

migration, and above all domestication are the primary causes of variation. At the beginning of his manuscript "Observations on Natural History," Hunter speculates in traditional terms about these "changes that have taken place in the productions of Nature by Time. . . ." He thinks that "we have reason to suppose there was a period in time in which every species of natural production was the same; there then being no variety in any species; but the variations taking place on the surface of the earth, such as the earth and water changing situations, which is obvious; as also the change in the poles or ecliptic, which I think is also obvious. The varieties [so produced] are but few and are still existing in what may be called the 'Natural' Animal. Also civilization has made varieties in many species, and without number, which are the 'Domesticated.'" *Essays,* 1: 3-4. Cf. ibid., 1: 37-39, 111-12, 240, 246, 308-12; *Observations,* pp. 277-85; *Lectures,* pp. 356-57. Obviously Hunter's belief that the "propagation or continuance of animals in their distinct classes is an established law of Nature" (*Observations,* p. 277) precluded the possibility of conceptualizing variation beyond the fixed units of original species.

58. Ibid., 1: 25-36. Hunter's considerations of taxonomy are quoted extensively and evaluated by Richard Owen in the prefaces to *Observations* (1837), pp. xxvii-xxxvi, and *Descriptive . . . Catalogue of the Physiological Series,* vol. 3 (1836). Hunter's failure to develop a theory of the "subordination of parts," and thus create a natural system of classification based on a physiologically deduced hierarchy of characters, is discussed below. Hunter would not have accepted the index championed with most success in the decades following his death—the nervous system. Cf. Georges Cuvier's classifactory *embranchements* of the animal kingdom, defined in terms of the four general plans of the nervous system, and echoed by Richard Owen, *Descriptive . . . Catalogue of the Physiological Series,* 3: vii, preface: "Now the nervous system may be said to be the essence of an animal, since all the other systems are subservient to and more or less indirectly influenced by it; and consequently it is found that the divisions of the animal kingdom which are characterized by the different types or plans according to which this system is disposed, are of primary value." John Hunter, however, as I explain in the second part of this essay, did not believe the nervous system to be thus physiologically dominant. The "internal animal oeconomy" in fact functions virtually autonomously according to his doctrine concerning this division of the organism. On the problem of identifying the characteristic or index organ in the systems of seventeenth- and eighteenth-century natural historians, see Phillip R. Sloan, "John Locke, John Ray, and the Problem of the Natural System," *Journal of the History of Biology* 5 (1972): 1-53; François Dagognet, *Le Catalogue de la vie: étude méthodologique sur la taxinomie* (Paris: Presses Universitaires de France, 1970), pp. 84-96.

59. *Essays,* 1: 145, and footnote: "The parts subservient to digestion in the complicated animals bear a great relation to the other properties of the animal. . . . Animals in general might be tolerably well classed by these organs, most being reducible to a few general classes, which again admit of many subdivisions." Hunter hints at the functional significance of the digestive system here, and indeed if any system is regarded as physiologically dominant by him, it is this.

60. Ibid., 1: 25, "On the Study of Natural History." Hunter's argument that Linnaeus has overlooked the anatomical complexity of the different types of mammal, a complexity that confuses his classificatory schema, is tremendously significant. Hunter rejects the general epistemological limits of eighteenth-century natural history, which characterized animals and plants by their surface features alone.

61. *Observations,* pp. 481-93, "Descriptions of Some Animals from New South Wales." Richard Owen mentions (p. 484, footnote) that Hunter also made notes on the

internal anatomy of these animals, but that Dr. George Shaw, editor of the Zoological Appendix to which Hunter contributed (see note 26) "was so blind to the true methods of advancing the science of zoology" that he did not include them. Here again is the contrast between natural history and John Hunter's problematic.

62. Ibid., pp. 485–86.

63. E.g., *Essays,* 1: 26–27: "The characters of the 'first class,' which includes land and sea animals, are: A heart made up of four cavities: essential. The lungs confined to a proper cavity, the enlargement of which is the cause of respiration: essential. Lungs divided into small cells: essential. Respiration quick: essential. Viviparous, and I believe the only animals that are so: essential. Give suck: essential. Parts of generation: in the male, made up of testes and one penis, the testes sometimes within and sometimes without the abdomen, but pass forwards; in the female, a clitoris, vagina, uterus, os uteri, fallopian tubes and ovaria: all essential. Kidneys high up in the abdomen: circumstantial. An external canal to the ear: circumstantial. Membrana typani concave externally: circumstantial. A cochlea: circumstantial. By much the most perfect animals, whether sea or land. There is a gradation from the land to the sea-animals, viz. Otter, Seal, Hippopotamus, Whale." The other classes described are Birds, "Lizards and Serpent-kind," and Amphibians ("more fishified than what the fish of the first class [Cetacea] are" [*sic*]).

64. At the end of the eighteenth century the chain of being concept came under attack from many anatomists and theorists of organic structure besides Hunter, notably from members of the Museum d'Histoire Naturelle in Paris, including J. B. Lamarck. See J. Schiller, "L'échelle des êtres et la série chez Lamarck," *Colloque international "Lamarck,"* ed. J. Schiller (Paris: Blanchard, 1971), pp. 87–103; M. J. S. Hodge, "Lamarck's Science of Living Bodies," *British Journal for the History of Science* 5 (1971): 329, 342–3. For an outline of a social history of the dissolution of the great chain, see William F. Bynum, "The Great Chain of Being after Forty Years: An Appraisal," *History of Science* 13 (1975): 1–28: also idem, "Time's Noblest Off-Spring: The Problem of Man in the British Natural Historical Sciences, 1800-1863," Ph.D. dissertation, University of Cambridge, 1974, pp. 1–42. Ironically, one of the latest affirmations of the concept came from a student of John Hunter, the Manchester surgeon Charles White (1728-1813). A lecture by Hunter on the varieties of human skull and the gradation of skulls from the human to the animal provided White, so he claimed, with the suggestion that there is a general gradation through the human races, from man through the animals, and from animals to vegetables: "Upon considering what Mr. Hunter thus demonstrated respecting skulls, it occurred to me that Nature would not employ gradation in one instance only, but would adopt it as a general principle." Charles White, *An Account of the Regular Gradation in Man and in Different Animals and Vegetables; and from the Former to the Latter* (London: C. Dilly, 1799), p. 41; cf. pp. iii, 142-44.

65. *Essays,* 1: 13.

66. Ibid., 1: 249.

67. Ibid., 1: 121.

68. *Treatise,* p. 184.

69. Ibid., p. 174.

70. *Essays,* 1: 384. Cf. the physiological series of the museum.

71. *Lectures,* p. 220. Hunter made similar use of comparative physiology; and he drew analogies from processes in plants as well as in animals. As he remarks at the beginning of his essay on the heat of animals and vegetables: "In the course of a variety of experiments on animals and vegetables, I have frequently observed that the result of

experiments in the one has explained the oeconomy of the other, and pointed out some principle common to both. . . ." John Hunter, "Of the Heat, &c. of Animals and Vegetables," *Philosophical Transactions of the Royal Society of London* 68 (1778): 7; reprinted in *Observations,* p. 136, footnote. An example of Hunter's use of such analogies is his belief that it is a law of nature that the "specific qualities in diseases," manifested, for example, by the formation of granulations, "tend more rapidly to the skin than to the deeper-seated parts. . . ." This is the principle, Hunter argues, by which plants always grow from the center of the earth toward the surface, and he interrupts his discussion of inflammation to describe some experiments on plant growth. *Treatise,* pp. 285–87, 489–90. Hunter believed that the fundamental physiological principles are common to both animals and plants. See *Essays,* 1:341–42; *Observations,* pp. 196–207. Two botanical manuscripts have survived: John Hunter, "Observations on Phytology," *Essays,* 1:341–68; idem, *Memoranda on Vegetation* (London: Taylor and Francis, 1860). For eighteenth-century uses of the "botanic analogy," see Philip C. Ritterbush, *Overtures to Biology: The Speculations of Eighteenth-Century Naturalists* (New Haven: Yale University Press, 1964).

72. See Part II of this essay, in which I discuss, to borrow Thomas Huxley's example in commenting upon this anatomical problematic, "the vastness of the interval which separates the physiology of John Hunter from the physiology of Johannes Muller and his successors." Thomas H. Huxley, "Owen's Position in the History of Anatomical Science," in Richard Owen, *The Life of Richard Owen by his Grandson* (London: John Murray, 1894; facs. reprint Farnborough: Gregg International, 1970), 2: 277. See Lloyd G. Stevenson, "Anatomical Reasoning in Physiological Thought," in *The Historical Development of Physiological Thought,* ed. Chandler McC. Brooks and Paul F. Cranefield. (New York: Hafner, 1959), pp. 27–28; Gerald M. Geison, "Social and Institutional Factors"; William F. Bynum, "The Anatomical Method, Natural Theology, and the Functions of the Brain," *Isis* 64 (1973): 445–68.

73. Perhaps the most influential comparative anatomist in this period to place major emphasis in constructing anatomical series upon functional considerations was Félix Vicq-d'Azyr. Cf. "Plan d'un cours d'anatomie et de physiologie," *Oeuvres de Vicq-d'Azyr,* 4: 37: "Les détails anatomiques, qui sont arides et rebutans par eux-mêmes, acquerront de l'intérêt, par les considérations que la Physiologie y mélera. . . . L'Anatomie seule n'est pour ainsi dire, que le squelette de la science; c'est la Physiologie qui lui donne du movement: l'une est l'étude de la vie, l'autre n'est que l'étude de la mort." In this course Vicq-d'Azyr drew up an unrealized program to compare the animal and plant organ systems which serve each of nine basic functions.

74. *Observations,* p. 482.

75. I am using the word "function" in its familiar contemporary sense. John Hunter, however, used the term synonymously with "action," referring to the use of an action simply by the word "purpose." I use the word "adapted" here rather generally, where Hunter might use alternatively "fitted for," "suited to," and other terms.

76. *Observations,* pp. 127–28.

77. *Essays,* 1:122.

78. Ibid., 1:145. In the case of the heart-lung system, Hunter summarizes the internal and external adaptations thus: "In those animals that have hearts we are to take into account a number of particulars. First, the blood's motion in consequence of that organ; secondly, the principal intention of that motion, viz. that it may be prepared in the lungs, which introduces breathing; thirdly, the variety in the kinds of lungs; fourthly, the different kinds of media in which the animals are obliged to breathe for the purpose of extracting matter employed in the preparation of this fluid [arterial blood]," *Treatise,*

p. 100. Cf. Vicq-d'Azyr's opinion that it behoves the astute comparative anatomist to "considérer avec soin et comparer ensemble deux espèces d'organes, dont les uns sont placés à la surface et les autres dans les grandes cavités. On peut regarder les premiers comme les instrumens immédiats de leurs mouvens [*sic*], et les seconds comme les ressorts cachés de la nutrition, de la sensibilité, de la reproduction et de la vie. Ces organes se correspondent; ils forment, en quelques sorte, les deux extrémités du système animal; et les uns ne peuvent éprouver de grands changemens, ni de grandes variétés, sans que les autres y participent." *Ouevres de Vicq-d'Azyr,* pp. 21-22. Vicq-d'Azyr illustrates this rule with relations between the organs of the digestive system, much as Hunter conceptualized them. Cf. Michel Foucault, *The Order of Things,* pp. 226-29, for an evaluation of this concept of organic structure constituted by anatomists such as Vicq-d'Azyr and Hunter in this period.

79. *Essays,* 1: 369-84.

80. Ibid., 1: 371.

81. Nerves are placed last for different reasons: "They are a good deal like the vessels in their dependencies; yet we choose to make them last, for these reasons: they are less understood, are more complicated, more numerous, and in general smaller." Ibid., 1: 372. Hunter generally argues that the internal organs and parts are very much more dependent on the blood than on the nerves; and he says very little about the anatomy and physiology of the nervous system. See "Description of the Nerves which Supply the Organ of Smelling," *Observations,* pp. 187-92; "Description of Some Branches of the Fifth Pair of Nerves," ibid., pp. 193-94; "Of the Brain and Nerves," *Essays,* 1:163-66.

82. Hunter's theory of the "internal animal oeconomy" is discussed below. Cf. *Observations,* p. 336: "I shall divide what is called the oeconomy of an animal: First, into those parts and actions which respect its internal functions, and upon which life immediately depends, as growth, waste, repair, shifting or changing of parts, &c., the organs of respiration and secretion, in which we include the powers of propagating the species. Secondly, into those parts and actions which respect external objects, and which are variously constructed, according to the kind of matter with which they are to be connected, whence they vary more than those of the first division. These are the parts for progressive motion, the organs of sense and the organs of digestion; all of which either act upon, or are acted upon by external matter."

83. *Lectures,* p. 225.

84. Georges Cuvier, "Tableau des rapports qui existent entre les variations des divers systêmes d'organes," *Leçons d'anatomie comparée,* 1: 45-60. I compare Hunter's ideas with the doctrine of Cuvier in greater detail below.

85. Richard Owen, *Essays,* 1: 125, footnote. The passage upon which Owen comments is one of Hunter's most opaque, but Owen's example seems to illustrate its sense.

86. Idem, *Observations,* "Preface," p. xxxiii.

87. Michel Adanson, *Familles des plantes* (1763), facs. reprint (Lehre: J. Cramer, 1966), presents 65 artificial arrangements of the genera of plants, each according to a single or very few characters. Adanson believed that his was the "natural method" of classification, in contrast to the "systems" that weighted characters a priori: "It was by looking at all those descriptions combined," he explained, "that I found that the plants arranged themselves naturally in classes or families that could not be systematic or arbitrary because they were not based on only one or a few parts which might change in some respects, but on all parts, so that the disappearance of one of those parts was compensated by the addition of another part." Evaluation of the affinities read from such an inventory of characters led the botanist to a knowledge of the genera. The above quotation is from Frans A. Stafleu, *Linnaeus and the Linnaeans: The Spreading*

of their Ideas in Systematic Botany, 1735-1789 (Utrecht: A. Oosthoek, 1971), p. 317. My interpretation of the contrast between "natural history" and "biology" follows Foucault, *The Order of Things*, pp. 141-45. See Vernon Pratt, "Foucault and the History of Classification Theory," *Studies in History and Philosophy of Science* 8 (1977): 163-71.

88. *Observations*, p. 336.

89. Foucault, *The Order of Things*, p. 231.

90. *Observations*, p. 422.

91. See Foucault, "La situation de Cuvier," pp. 73-75. For a history of nineteenth-century comparative anatomy that takes the conflict between transcendentalism and functionalism in morphology as its interpretative viewpoint, see E. S. Russell, *Form and Function: A Contribution to the History of Animal Morphology* (London: John Murray, 1916). Richard Owen's work is best understood in terms of a tension between these two approaches within it; see Roy M. MacLeod, "Evolutionism and Richard Owen, 1830-1868: An Episode of Darwin's Century," *Isis* 56 (1965): 259-80. The Hunterian Museum—an unbound text, ever plastic in its content and grammar—demonstrates these transformations in morphological and classificatory style in comparative anatomy. William Henry Flower (1831-1899), Conservator and Hunterian Professor at the Royal College of Surgeons, who felt it was his duty to the specimens in the collection, as "*their* mouthpiece . . . to endeavour to put their silent eloquence in some sort of articulate language," observed in 1881: "As the collection continues to advance, the classification according to homology is gradually superseding that according to analogy, with which it began." William Henry Flower, *Essays on Museums and Other Subjects Connected with Natural History* (London: Macmillan, 1898), pp. 86, 97-98. Not surprisingly, under Flower's direction the museum grew above all in the division of comparative osteology. See Jessie Dobson, "Conservators of the Hunterian Museum IV, William Henry Flower," *Annals of the Royal College of Surgeons of England* 30 (1962): 383-91.

92. Foucault, "La situation de Cuvier," pp. 69, 86.

93. Cf. *Essays*, 1: 121.

94. Cf. ibid., 1: 122, 145, and see note 78.

95. Cf. *Essays*, 1: 145.

96. Georges Cuvier, *Le Règne animal distribué d'après son organisation pour servir de base à l'histoire naturelle des animaux et d'introduction à l'anatomie comparée*, 4 vols. (Paris: Déterville, 1817). See also idem, *Histoire des sciences naturelles*, "Des méthodes," pp. 13-48; Henri Daudin, *Cuvier et Lamarck*, 2: 13-68; William Coleman, *Georges Cuvier*, pp. 74-106. The concept of a subordination of characters had been introduced first in botany by Antoine-Laurent de Jussieu (1748-1836). Here, however, the order of characters was derived by the method of calculating their relative frequencies; Cuvier's concept of subordination was the product of physiological theory, and had the status of an a priori principle in classification. See Coleman, pp. 102-4; Foucault, *The Order of Things*, p. 263.

97. The library of the Royal College of Surgeons of England possesses four volumes of John Hunter's case books (William Clift's transcript): "Cases in Surgery," "Acounts of the Dissections of Morbid Bodies," "Cases and Observations," "Cases and Dissections." These contain over 700 different records of cases; see Jessie Dobson, "John Hunter's Practice," *Annals of the Royal College of Surgeons of England* 38 (1966): 181-90. Hunter's manuscript catalogue of the pathological specimens in his museum, in Clift's hand, is also preserved at the Royal College of Surgeons. See *List of Books, Manuscripts, Portraits, etc., Relating to John Hunter in the Royal College of Surgeons of England* (London: Taylor and Francis, 1891). I have not inspected this material, but much of it

has been published in catalogues of the pathological collection. The first truly systematic British treatise of the whole of morbid anatomy was written by Hunter's nephew and protégé, Matthew Baillie (1761-1823). This work was based largely on William Hunter's anatomical collection, to which John contributed. Matthew Baillie, *The Morbid Anatomy of Some of the Most Important Parts of the Human Body* (London: J. Johnson and G. Nicol, 1793); reprinted from the second American edition (1808) in Alvin E. Rodin, *The Influence of Matthew Baillie's Morbid Anatomy* (Springfield, Ill.: Charles C. Thomas, 1973). Hunter represented a trend toward the end of the eighteenth century for surgeons to play a leading role in the development of medicine, above all through pathological anatomy, but also professionally in private and hospital practice; indeed, to demonstrate that surgeons were of influence, "the name of John Hunter alone would suffice." Owsei Temkin, "The Role of Surgery in the Rise of Modern Medical Thought," *Bulletin of the History of Medicine* 25 (1951): 487.

98. For concise discriminations between ontological and physiological conceptions of disease in medical history see Lelland J. Rather, trans. and ed., *Disease, Life and Man: Selected Essays by Rudolf Virchow* (Stanford: Stanford University Press, 1958), "Introduction," pp. 17-20; Peter H. Niebyl, "Sennert, van Helmont, and Medical Ontology," *Bulletin of the History of Medicine* 45 (1971): 115-20. On eighteenth-century nosology see Vaidy, "Nosographie," *Dictionaire des sciences médicales* (Paris: C.L.F. Panckoucke, 1819), 36: 206-65; Knud Faber, *Nosography: The Evolution of Clinical Medicine in Modern Times* (New York: Paul B. Hoeber, 1930), pp. 1-58; Lester S. King, *The Medical World of the Eighteenth Century* (Chicago: University of Chicago Press, 1958), pp. 193-226.

99. *Lectures,* p. 130. The symptoms to which Hunter gives most attention are varieties of specific, local, inflammatory lesions.

100. For interpretations of transformations in the meaning of signs and symptoms in this period see Georges Gusdorf, "La constitution d'un champ épistémologique: la nosologie," in *Dieu, la nature, l'homme au siècle des lumières* (Paris: Payot, 1972), pp. 477-98, especially pp. 495-98; Alain Rousseau, "Une révolution dans la sémiologie médicale: le concept de spécificité lésionnelle," *Clio Medica* 5 (1970): 123-31; Karl Figlio, "The Historiography of Scientific Medicine: An Invitation to the Human Sciences," *Comparative Studies in Society and History* 19 (1977): 277-81. John Hunter's physiological concept of "vital property," and its structural symmetry with his concept of "disposition to disease," are discussed in the second part of this paper; the ultimate reference of both his physiological and pathological theories was to neither actions nor processes, but the properties–or powers–of parts. My interpretation of the isomorphism between natural history and ontological medicine, and between physiological pathology and biology, draws especially upon the general thesis of Michel Foucault, *The Birth of the Clinic: An Archaeology of Medical Perception,* trans. A. M. Sheridan Smith (New York: Vintage Books, 1975).

101. See François Duchesneau, "La philosophie médicale de Sydenham," *Dialogue* 9 (1970): 54-68; idem, *L'Empirisme de Locke* (The Hague: Martinus Nijhoff, 1973), pp. 1-41; Lester S. King, "Boissier de Sauvages and 18th Century Nosology," *Bulletin of the History of Medicine* 40 (1966): 43-51. On the botanical aspect of eighteenth-century nosology, see Frederik Berg, "Linné et Sauvages. Les rapports entre leurs systèmes nosologiques," *Lychnos* (1956): 31-54; François Dagognet, *Le Catalogue de la vie,* pp. 125-71. For an example of the use of the Galenic concept of disease as *functio laesa* in the early eighteenth century (Boerhaave) see Owsei Temkin, "The Scientific Approach to Disease: Specific Entity and Individual Sickness," in *Scientific Change,* ed. A. C. Crombie (New York: Basic Books, 1963), p. 637.

102. William Cullen, *First Lines of the Practice of Physic* (Edinburgh: C. Elliot, 1784), 3: 121-3 (sections 1090-92). "In a certain view, almost the whole of the diseases of the human body might be called Nervous. . . ." Cf., however, idem, *Nosology: Or, a Systematic Arrangement of Diseases,* trans. anon. (Edinburgh: Bell and Bradfute, 1810), "Preface," pp. xvii-xviii: "In the first place, therefore, I have always selected those external marks which are easily observable by our senses; neglecting, or rather rejecting any conjectures as to the internal state of the body. . . . In the second place, I have considered those symptoms as affording the chief characteristic marks, which are the never failing attendants of the disease. . . . Ought the cause of a disease to make any part of the definition? To this it may be answered, that as the judgment formed by physicians of the causes of diseases, is often fallacious, and even false, and therefore not to be rashly relied on in distinguishing diseases; yet, as these causes are sometimes sufficiently certain, and *easily to be observed,* they may be admitted in Nosology, as *legitimate characters.*" (Emphasis added.) See Inci Altug Bowman, "William Cullen (1710-90) and the Primacy of the Nervous System," Ph.D. dissertation, Indiana University, 1975, pp. 153-203; John Thomson, *An Account of the Life, Lectures, and Writings of William Cullen, M.D.* (Edinburgh: William Blackwood, 1859), 2: 1-77.

103. John Brown, *The Elements of Medicine,* in *The Works of Dr. John Brown,* ed. William Cullen Brown (London: J. Johnson, 1804), vol. 2; Erasmus Darwin, *Zoonomia; or the Laws of Organic Life,* 2 vols. (London: J. Johnson, 1794-1796). The functional basis of these systems of classification is emphasized in W. Riese, "History and Principles of Classification of Nervous Diseases," *Bulletin of the History of Medicine* 18 (1945): 468-80. See also Gernot Rath, "Neural Pathology: A Pathogenetic Concept of the 18th and 19th Centuries," *Bulletin of the History of Medicine* 33 (1959): 526-41; idem, "Die Neuropathologie am Ausgang des 18. Jahrhunderts," in *Von Boerhaave bis Berger,* ed. K. E. Rothschuh (Stuttgart: Gustav Fisher, 1964), pp. 35-47; Gunter B. Risse, "The Brownian System of Medicine: Its Theoretical and Practical Implications," *Clio Medica* 5 (1970): 45-51; R. R. Trail, "Sydenham's Impact on English Medicine," *Medical History* 9 (1965): 363.

104. *Lectures,* p. 211, original emphasis.

105. Ibid., p. 219. John Abernethy (1764-1831), a staunch Hunterian advocate in the early nineteenth century, no doubt expressed a general professional sentiment when he viewed his master as "a new character in our profession; and briefly to express his peculiar merit, I may call him the first and great physionosologist or expositor of the nature of disease." John Abernethy, *The Hunterian Oration for the Year 1819* (London: Longman, Hurst, Rees, Orme, and Brown, 1819). Whatever the intellectual consequences, John Hunter's mature professional life was characterized by his struggle to illuminate the practices of surgery and medicine with science. In contrast to the positive assessment of the following generation, an ideological product of the original members of the newly chartered Royal College of Surgeons, Hunter's scientific and related pedagogical interests met with resentment and resistance in his peers. For examples of this see Jessé Foot, *The Life of John Hunter,* pp. 84-85, 263-64; Bernice Hamilton, "The Medical Profession in the Eighteenth Century," *Economic History Review,* second series, 4 (1951): 155; Lloyd A. Wells, "The Intractability of One Man: Hunterian Controversies," *Perspectives in Biology and Medicine* 19 (1976): 315-22. For a suggestive theory by which the "recurrent" readings of John Hunter—as the "first of the moderns"—might be contextualized, see Judy Sadler, "Ideologies of 'Art' and 'Science' in Medicine: The Transition from Medical Care to the Application of Technique in the British Medical Profession," in *The Dynamics of Science and Technology: Social Values, Technical Norms, and Scientific Criteria in the Development of Knowledge,* ed. Wolfgang Krohn et al. (Dordrecht:

D. Reidel, 1978), pp. 177–215. A well-considered assessment of the scientific character of Hunter's approach to innovation in surgery and medicine is Lloyd Allan Wells, "Aneurysm and Physiologic Surgery," *Bulletin of the History of Medicine* 44 (1970): 411–24. For a survey of relations between physiology and pathology see Joseph Schiller, "The Influence of Physiology on Medicine," *Episteme* 6 (1972): 116–27. My discussion of Hunter's pathological doctrines is confined strictly to the level of theory; it should be stressed that the (reciprocal) relations between theory and practice in medicine may be very highly mediated indeed by extra-scientific factors. See Charles E. Rosenberg, "The Therapeutic Revolution: Medicine, Meaning and Social Change in Nineteenth Century America," *Perspectives in Biology and Medicine* 20 (1977): 485–506.

106. *Observations*, p. 117.

107. *Lectures*, p. 225.

108. *Observations*, p. 168.

109. Cf. Xavier Bichat, *Recherches physiologiques sur la vie et la mort* (Paris: Gabon, 1800), pp. 163–372. See Foucault, *The Birth of the Clinic*, pp. 124–48, especially pp. 140–46; and M. D. Grmek, "Changements historiques du concept de mort," *Proceedings of the 13th International Congress of the History of Science* (Moscow: Nauka, 1974), 9: 11–21.

110. *Essays*, 1: 239.

111. Ibid., 1: 240. Cf. *Observations*, p. 44: "Every deviation from that original form and structure which gives the distinguishing character to the productions of Nature, may not improperly be called monstrous. According to this acceptation of the term, the variety of monsters will be almost infinite; and, as far as my knowledge has extended, there is not a species of animal, nay, there is not a single part of an animal body, which is not subject to an extraordinary formation." Hunter wrote two papers on hermaphrodites: "Account of the Free-Martin," *Observations*, pp. 34–43; "Account of an Extraordinary Pheasant," ibid., pp. 44–49; and a set of notes "On Monsters," *Essays*, 1: 239–51.

112. *Essays*, 1: 249, and following quotations.

113. John Hunter's discourse reveals already the *normative* quality of the concept of life as it emerged in the nineteenth century; for a discussion of this conceptual structure—life/anomaly—see Georges Canguilhem, *On the Normal and the Pathological*, trans. Carolyn R. Fawcett (Dordrecht: D. Reidel, 1978), pp. 69–85. For a general discussion of the history of teratology, see idem, *La Connaissance de la vie* (Paris: J. Vrin, 1975), pp. 171–84. The possibility of a rational science of monstrous forms engendered by the concept of organic structure, was realized in the early nineteenth century above all by Étienne Geoffroy Saint-Hilaire (1772–1844). See Théophile Cahn, *La Vie et l'oeuvre d'Étienne Geoffroy Saint-Hilaire* (Paris: Presses Universitaires de France, 1962), pp. 167–85.

114. *Lectures*, p. 220. Cf. the position of Cuvier outlined in William Coleman, "Les organismes marins et l'anatomie comparée dite expérimentale: l'oeuvre de Georges Cuvier," in *Colloque international sur l'histoire de la biologie marine, Vie et milieu*, supplément 19 (Paris: Mason, 1965), pp. 225–38.

115. See Brigitte Hoppe, "Le concept de biologie chez G. R. Treviranus," in *Colloque international "Lamarck,"* pp. 199–237; J. Schiller, "A propos de la diffusion du terme biologie," in ibid., pp. 239–42. The word "biology" was introduced into English by William Lawrence in his lectures at the Royal College of Surgeons in 1817; see William Lawrence, *Lectures on Physiology, Zoology and the Natural History of Man* (London: J. Callow, 1819), p. 60. Cf. ibid., pp. 63–64: "Thus we find that anatomy, physiology, morbid anatomy, and pathology are mutually related and intimately connected. Although called separate sciences, they are, in truth, parts of one system. . . . The collection

of this college was formed, and is now arranged, in conformity to the views just alluded to. . . . Mr. Hunter was the first in this country who investigated disease in a strictly physiological method. . . ."

116. Hunter defines "life, in which consists the power of self-preservation," as the *antidote* to "decay, or spontaneous reduction to common matter, when life is gone." *Lectures,* p. 215. Cf. the better known formula of Xavier Bichat (1771–1802) in *Recherches physiologiques,* p. 1: "La vie est l'ensemble des fonctions qui resistent à la mort." Many aspects of Hunterian doctrine in the following pages will bear close comparison with opinions of Bichat, as also with the Montpellier school of medical vitalism to which Bichat's work owed considerable debts. See Elizabeth Haigh, "The Roots of the Vitalism of Xavier Bichat," *Bulletin of the History of Medicine* 49 (1975): 72–86, especially pp. 73–79, for a discussion of this last point. The Montpellier tradition has often been regarded as a revision of the animism of George Ernst Stahl (1659–1734); notably, Stahl too defined life dialectically as the active living principle that conserves the body, resisting putrefaction and corruption during life. Stahl's *anima,* however, as I discuss below, presupposes a theory of vital action significantly different from that developed by the Montpellier physicians and by John Hunter in the second half of the eighteenth century. See note 129 below. Hunter's vital principle should not, therefore, be interpreted as the same concept as that of the neo-Stahlian Montpellier physician Paul-Joseph Barthez (1734–1806), which it recalls. See Réjane Bernier, "La notion de principe vital chez Barthez," *Archives de philosophie* 35 (1972): 423–41; Elizabeth L. Haigh, "The Vital Principle of Paul Joseph Barthez: The Clash between Monism and Dualism," *Medical History* 21 (1977): 1–14.

117. *Treatise,* p. 117.

118. Ibid., p. 106.

119. Ibid., p. 107.

120. *Essays,* 1: 117. Hunter believed also that different parts of the bodies of the structurally more complex animals are possessed of the living power in different degrees. This is demonstrated, in his opinion, by a number of experiments on the transplantation of spurs and testes in cocks and hens, and by his routine surgical practice of transplanting human teeth. John Hunter, *The Natural History of the Human Teeth,* in *Works,* 1: 55. See also *Essays,* 1: 19; *Treatise,* pp. 273–75. For a discussion of this point see C. Barker Jørgensen, *John Hunter, A. A. Berthold, and the Origins of Endocrinology* (Odense: Odense University Press, 1971), pp. 17–20.

121. *Essays,* 1: 147.

122. *Lectures,* p. 242. Hunter continues, "for according to this definition, a dead body is as much organized as a living one, for in the dead body the same *mechanism* exists as in the living one." (Emphasis added.) The *effects* of vital actions in the living animal machine are as mechanical as the actions of the dead body—stretching of elastic tendons, for example. I discuss Hunter's meaning of the term "animal machine" below.

123. *Lectures,* p. 214.

124. Ibid., p. 216. Cf. *Treatise,* p. 103. According to Theodore Brown, John Hunter was responsible for reversing the decline of English physiological research in the 1770's by dismantling the Royal Society's mechanistic physiology and producing a "phenomenalist vitalism." Theodore M. Brown, "From Mechanism to Vitalism," pp. 181, 192–93. On the decline of iatromechanism see also Robert Schofield, *Mechanism and Materialism: British Natural Philosophy in an Age of Reason* (Princeton: Princeton University Press, 1970), pp. 191–209. For a general discussion of vitalist arguments for the insufficiency of iatromechanical models see François Duchesneau, "Malpighi, Descartes, and the Epistemological Problems of Iatromechanism," in *Reason, Experiment and*

Mysticism in the Scientific Revolution, ed. M. L. Righini Bonelli and William R. Shea (New York: Science History Publications, 1975), pp. 111–30.

125. *Lectures,* pp. 219–20.

126. *Essays,* 1: 4–18. On "active principles" in eighteenth-century natural philosophy, see P. M. Heimann and J. E. McGuire, "Newtonian Forces and Lockean Powers: Concepts of Matter in Eighteenth-Century Thought," *Historical Studies in the Physical Sciences* 3 (1971): 233–306; J. E. McGuire, "Force, Active Principles and Newton's Invisible Realm," *Ambix* 15 (1968): 154–208; Arnold Thackray, "'Matter in a Nut-Shell': Newton's *Opticks* and Eighteenth-Century Chemistry," *Ambix* 15 (1968): 29–53. The Edinburgh tradition in natural philosophy as expressed by William Cullen, with which John Hunter no doubt had some familiarity, is discussed in A. L. Donovan, *Philosophical Chemistry in the Scottish Enlightenment: The Doctrines and Discoveries of William Cullen and Joseph Black* (Edinburgh: Edinburgh University Press, 1975), pp. 19–162.

127. *Lectures,* p. 222.

128. Ibid.

129. *Essays,* 1: 7–8.

130. Georges Cuvier, *Rapport historique,* p. 230; Georges Canguilhem, *La Connaissance de la vie,* p. 156. Stahl's physiological conceptions may be examined most conveniently in George Ernst Stahl, "Über den Unterschied zwischen Organismus und Mechanismus" (1714), in *Sudhoffs Klassiker der Medizin,* Band 36, trans. from Latin B. J. Gottlieb, ed. and intro. Rudolph Zaunick (Leipzig: Johann A. Barth, 1961), pp. 48–53. The historiography outlined in this paragraph is developed in Paul Delaunay, "L'évolution philosophique et médicale du biomécanicisme. De Descartes à Boerhaave, de Leibnitz à Cabanis," *Le progrès médical* 34 (1927): 1289–93 and 35: 1337–52; 36: 1369–84; Owsei Temkin, "The Philosophical Background of Magendie's Physiology," *Bulletin of the History of Medicine* 20 (1946): 10–35; idem, "Materialism in French and German Physiology of the Early Nineteenth Century," Bulletin of the History of Medicine 20 (1946): 322–27; Richard Toellner, "Anima et Irritabilitas. Hallers Abwehr von Animismus und Materialismus," *Sudhoffs Archiv* 51 (1967): 130–44; idem, *Albrecht von Haller: über die Einheit im Denken des letzten Universalgelehrten, Sudhoffs Archiv,* Beiheft 10 (Weisbaden: Franz Steiner, 1971), pp. 171–82; Joseph Schiller, "Queries, Answers, and Unsolved Problems in Eighteenth-Century Biology," *History of Science* 12 (1974): 184–99, especially 192–94; Karl M. Figlio, "The Metaphor of Organization," pp. 25–26; Elizabeth L. Haigh, "The Vital Principle of Paul Joseph Barthez," pp. 1–8. On Stahl see François Duchesneau, "G. E. Stahl: antimécanisme et physiologie," *Archives internationales d'histoire des sciences* 26 (1976): 3–26; L. J. Rather, "G. E. Stahl's Psychological Physiology," *Bulletin of the History of Medicine* 35 (1961): 37–49. On Hoffmann's critique of Stahl see Lester S. King, "Stahl and Hoffmann: A Study of Eighteenth Century Animism," *Journal of the History of Medicine* 19 (1964): 118–30. General discussions of the animist-mechanist debate may be found in R. K. French, "Sauvages, Whytt and the Motion of the Heart: Aspects of Eighteenth Century Animism," *Clio Medica* 7 (1972): 35–54; Lester S. King, "Basic Concepts of Early 18th-Century Animism," *American Journal of Psychiatry* 124 (1967): 797–802. My interpretation of John Hunter's conception of living matter, while seldom made in recent scholarship, is by no means original. William Lawrence, for example, utilized similar themes for his own purposes. Later in the nineteenth century advocates of biological materialism could appeal to a "Hunterian stage" which was "said to form the second or intermediate stage of opinion, viz.–the incomplete separability of matter and force, or life; the third, or modern idea, being the complete inseparability of matter and force." Henry I. Fotherby,

Scientific Associations, their Rise, Progress, and Influence, with a History of the Hunterian Society (London: Bell and Daldy, 1869), p. 53.

131. *Lectures,* pp. 214-15.

132. Hunter's concept of "animalization" refers to an unexplained process whereby matter is endowed with the "arrangement" of life, and should not be confused with chemical theories of the transformation of the elementary composition of nutrients in digestion, and of the blood in respiration, such as were developed by Antoine de Fourcroy, Antoine Lavoisier, or Jean Noël Hallé. See Frederic L. Holmes, "Elementary Analysis and the Origins of Physiological Chemistry," *Isis* 54 (1963): 50-81; D. C. Goodman, "The Application of Chemical Criteria to Biological Classifications in the Eighteenth Century," *Medical History* 15 (1971): 29-34.

133. *Lectures,* p. 231. Cf. *Essays,* 1: 13. Hunter believed that "whatever changes, processes, additions, or reductions" the nutritive juice may undergo in the lungs are not yet known. *Essays,* 1: 16. See note 203 below.

134. *Lectures,* p. 223.

135. Ibid., pp. 222-23.

136. Ibid., p. 223.

137. *Treatise,* p. 176.

138. *Lectures,* pp. 242-43.

139. *Treatise,* pp. 150-55.

140. *Lectures,* p. 219, emphasis added. The concept of an organic machine most readily brings to mind René Descartes' (1596-1650) model of the body as a mechanical automaton in his *Treatise of Man* (French ed. 1664), trans. and commentary Thomas Steele Hall (Cambridge: Harvard University Press, 1972). It is, however, to be found in the doctrines of animists and anti-Cartesian vitalists from the middle of the seventeenth century, the beginning of the period Michel Foucault has termed the classical age. See Joseph Schiller, "La notion d'organisation dans l'oeuvre de Louis Bourguet (1678-1742)," *Gesnerus* 32 (1975): 87-97. Schiller has claimed that in the eighteenth century the synonymous term and the concept of "organization" "are adopted by every single naturalist. . . ." Idem, "Queries, Answers and Unsolved Problems," p. 92, footnote 32. See also references in Phillip R. Sloan, "Descartes, the Sceptics, and the Rejection of Vitalism in Seventeenth-Century Physiology," *Studies in History and Philosophy of Science* 8 (1977): 3, footnote 4. For a discussion of late eighteenth-century vitalist theories of *l'homme machine* as a living machine, see Aram Vartanian, "Cabanis and La Mettrie," *Studies on Voltaire and the Eighteenth Century* 155 (1976): 2149-66; idem, *La Mettrie's L'Homme Machine: A Study in the Origins of an Idea* (Princeton: Princeton University Press, 1960), pp. 114-29; Sergio Moravia, "From *homme machine* to *homme sensible:* Changing Eighteenth Century Models of Man's Image," *Journal of the History of Ideas* 39 (1978): 45-60. See also G. Vigarello, "Buffon et la 'machine animale'," *Episteme* (1973): 186-98.

141. Albrecht von Haller, "A Dissertation on the Sensible and Irritable Parts of the Body" (London: 1755), trans. anon., ed. Owsei Temkin, *Bulletin of the History of Medicine* 4 (1936): 692. For a discussion of Haller's "Newtonian" conception of vital force, see Shirley A. Roe, "The Development of Albrecht von Haller's Views on Embryology," *Journal of the History of Biology* 8 (1975): 180-84. Cf. Xavier Bichat, *Anatomie générale, appliquée à la physiologie et à la médecine* (Paris: Brosson, Gabon, 1801), pp. viii, xxxvii, translated and quoted by Pedro Lain Entralgo, "Sensualism and Vitalism in Bichat's 'Anatomie générale,'" *Journal of the History of Medicine* 3 (1948): 52-53: "The relationship between properties as causes, and phenomena as effects, is an axiom whose repetition has today become almost irritatingly familiar in physics,

chemistry, astronomy, and so forth. If this work establishes an analogous axiom in physiological science, it will have achieved its goal. . . . Physical science, equally with physiological, is found to be composed of two things: first, the study of phenomena, which are the effects, and secondly, examination of the connexions subsisting between these and physical or vital properties, which are the causes." On Bichat's notion of physiological force see E. Benton, "Vitalism in Nineteenth-Century Physiological Thought: A Typology and Reassessment," *Studies in History and Philosophy of Science* 5 (1974): 25-29.

142. *Essays,* 1: 152. Hunter argues here that absorption at the mouths of the lymphatic vessels "is certainly not begun according to the principle of attraction by capillary tube"–i.e., the mechanical effect of capillarity. These parts have a special power of action, analogous to "drinking." Hunter discusses the mechanical non-proportionality of impulse and response most explicitly in *Essays,* 1: 341. Cf. Schiller, "Queries, Answers, and Unsolved Problems," p. 196, footnote 11: "Irritability should be conceived as the intermediary agent between the external environment (the stimulus) and the organism which responds by motion, secretion, choice of the aliment, depending on the organ involved. . . . Inside the organism an organ is external to another organ which may be irritated by it." A notable illustration of this concept of internal irritation is the stimulation of muscle by nerve. Haller, in particular, popularized the notion that nerve force simply activates the muscle fiber's own *vis insita.* See Stanley W. Jackson, "Force and Kindred Notions in Eighteenth-Century Neurophysiology and Medical Psychology," *Bulletin of the History of Medicine* 44 (1970): 549-54.

143. *Essays,* 1: 164. Cf. Albrecht von Haller, "A Dissertation on the Sensible and Irritable Parts," p. 658. To his nineteenth-century eulogists John Hunter was physiology's Newton. Of the several early Hunterian Orators, William Blizzard's comments are typical: "His opinions, and doctrines, were deductions from manifold observations, experiments, and inquiries; not assumptions, framed into hypotheses, and asserted by appropriate jargon." William Blizzard, *The Hunterian Oration Delivered in the Theatre of the Royal College of Surgeons* (London: Rivingtons, 1815), p. 72. The monistic, vitalist alternative to the dualisms of Descartes and Stahl, has been situated, in accord with rhetoric easily found in eighteenth-century physiological texts, in the "Newtonian" tradition of scientific explanation. J. Goodfield-Toulmin, "Some Aspects of English Physiology: 1780-1840," *Journal of the History of Biology* 2 (1969): 283-320, discusses the concept of properties of animal matter in John Hunter and his followers in terms of "Newtonian phenomenalism." This is the theme of a wider-ranging survey, Thomas S. Hall, "On Biological Analogs of Newtonian Paradigms," *Philosophy of Science* 35 (1968): 6-27. The existence of an "underground stream" of faculty- or property–based physiology flowing from Newtonian debates of the first decades of the eighteenth century is demonstrated in Theodore M. Brown, "From Mechanism to Vitalism," pp. 185-89.

144. My argument demonstrates, in support of the historiographical perceptions of Foucault, Albury, and others, that in Hunter the relations of function and structure are conceptualized in a manner radically different from that within the functionalist problematic of nineteenth-century biology. I do not, therefore, endorse the opinion of Joseph Schiller that "organization" as conceptualized by eighteenth-century natural historians "was the fundamental pre-condition for the elaboration of the concept of biology at the beginning of the nineteenth century." Schiller, "Queries, Answers, and Unsolved Problems," p. 189. Cf. ibid., p. 194; idem, "La Notion d'organisation," pp. 92-93; idem, *La Notion d'organisation dans l'histoire de la biologie* (Paris: Maloine, 1978), passim. An overview of theories–both vitalist and mechanist–of fiber properties in the

eighteenth century may be gained from Albrecht von Haller, "A Dissertation on the sensible and irritable parts," pp. 692-96; ibid., editor's introduction, pp. 651-55; Aram Vartanian, *La Mettrie's L'Homme Machine*, pp. 82-89, 234-36; L. J. Rather, "Some Relations between Eighteenth-Century Fiber Theory and Nineteenth-Century Cell Theory," *Clio Medica* 4 (1969): 191-202; M. D. Grmek, "La notion de fibre vivante chez les médecins de l'école iatrophysique," *Clio Medica* 5 (1970): 297-318; E. Bastholm, *The History of Muscle Physiology from the Natural Philosophers to Albrecht von Haller* (København: Ejnar Munksgaard, 1950); Thomas S. Hall, *History of General Physiology 600 B.C. to A.D. 1900*, 2 vols. (Chicago: University of Chicago Press, 1975).

145. *Treatise*, p. 106.

146. Ibid., p. 2.

147. *Observations*, p. 197.

148. E.g., ref., note 141. Since Théophile Bordeu, *Recherches anatomiques sur la position des glandes et leur action* (1751) in *Oeuvres complètes de Bordeu*, ed. B.-A. Richerand (Paris: Caille, Ravier, 1818), 1: 49-208, one of the manifestos of late eighteenth-century vitalism, the vital property of glands was a model of sensibility. See Elizabeth L. Haigh, "Vitalism, the Soul, and Sensibility: The Physiology of Théophile Bordeu," *Journal of the History of Medicine* 31 (1976): 30-41. Cf. Erasmus Darwin, *Zoonomia*, 1: 493: "such varieties of irritability or of sensibility exist in our adult stage in the glands; every one of which is furnished with an irritability, or a taste, or appetency, and a consequent mode of action peculiar to itself."

149. Pierre-Jean-Georges Cabanis, *Rapports du physique et du moral de l'homme* (1802), in *Oeuvres complètes de Cabanis*, (Paris: Firmin Didot, 1824) 3: 105, translated and quoted by William Randall Albury, "Experiment and Explanation," p. 86. The illustration of the statue enjoyed its most famous application in Étienne Bonnot, abbé de Condillac, *Traité des sensations* (1754), in *Oeuvres philosophiques de Condillac*, ed. Georges LeRoy (Paris: Presses Universitaires de France, 1947-1951), 1: 119-319. Throughout the eighteenth century John Locke, *An Essay Concerning Human Understanding* (1690) remained the foundation stone of empiricist epistemology. In commenting upon the passage cited from Cabanis, Karl Figlio has argued that toward the end of the eighteenth century sensibility, rooted in sensualist philosophy, came to subsume conceptions of irritable response. The Hallerian distinction between irritability and sensibility was replaced by the notion that all physiological effects may be attributed to the sensible power of living nerve. Karl M. Figlio, "The Metaphor of Organization," p. 22. See also idem, "Theories of Perception and the Physiology of Mind in the Late Eighteenth Century," *History of Science* 12 (1975): 177-212. Similarly, Michael Gross has argued that at the turn of the century local sensibility was increasingly replaced by a discrete faculty localized in the nervous system: Michael Gross, "Function and Structure in Nineteenth Century French Physiology," Ph.D. dissertation, Princeton University, 1974, pp. 12-74; idem, "The Lessened Locus of Feelings: A Transformation in French Physiology in the Early Nineteenth Century," *Journal of the History of Biology* 12 (1979): 231-71. It is from this position that Georges Cuvier, following Cabanis himself, characterized the Montpellier vitalists Bordeu and Barthez, Erasmus Darwin in England, Cabanis, and Blumenbach as "neo-stahlians," who, he commented critically, "attribuaient . . . l'irritabilité à une sensibilité, mais à une sensibilité locale, différente pour chaque organe." Georges Cuvier, *Histoire des sciences naturelle*, 1: 330. In John Hunter sensibility subsumes irritability, but remains local. Furthermore, local sensibility, for Hunter, as I shall demonstrate below in detail, does not depend upon the sensitivity and action of nerves, nor, indeed, upon nervous connection with the sensible part. For extended discussions of Montpellier vitalism, see Martin S. Staum, "Cabanis and the

Science of Man," Ph.D. dissertation, Cornell University, 1971; idem, "Medical Components in Cabanis's Science of Man," *Studies in History of Biology* 2 (1978): 1–32; Elizabeth Haigh, "Roots of the Vitalism of Xavier Bichat," Ph.D. dissertation, University of Wisconsin, 1973; Jacques Roget, *Les sciences de la vie dans la pensée française du XVIIIe siècle* (n. p.: Armand Colin, 1963), pp. 614–54. The rise of sensibility in general intellectual culture is sketched in G. S. Rousseau, "Nerves, Spirits, and Fibres: Towards Defining the Origins of Sensibility," *Studies in the Eighteenth Century,* ed. R. F. Brissenden and J. C. Eade (Toronto: University of Toronto Press, 1976), 3: 137–57.

150. *Essays,* 1: 133.

151. Ibid., 1: 115.

152. *Treatise,* p. 116. Hunter introduces the hypothesis of a *materia vitae diffusa* to explain this mechanism; see below.

153. *Lectures,* pp. 274–75. "Custom is with me the negative of habit: by custom comes an insensibility to impression, the impression diminishing although the cause is the same, and the parts becoming more and more at rest." Cf. p. 277, where such physiological effects, mental intransigence, and the "*vis inertiae* in matter" are all attributed to the principle referred to as habit.

154. Ibid., p. 236, original emphasis.

155. See William Randall Albury, "Experiment and Explanation," pp. 75–80. The most comprehensive study that I have seen of the *Idéologues,* science and language is idem, "The Logic of Condillac and the Structure of French Chemical and Biological Theory, 1780–1801," Ph.D. dissertation, Johns Hopkins University, 1972. See also Sergio Moravia, "Philosophie et médecine en France à la fin du XVIIIe siècle," *Studies on Voltaire and the Eighteenth Century* 89 (1972): 1138–51.

156. *Essays,* 1: 120.

157. Idiosyncracies of terminology and other problems with language have no doubt contributed largely to the failure of Hunterian exegetes to recognize the typically eighteenth-century structure of John Hunter's vitalism. The fact that the concepts of irritability and sensibility are expressed by Hunter in unfamiliar terms is taken here as support for the general thesis of this paper: namely, that beneath the explicit level of language and theoretical nuance in eighteenth-century physiology, there may be found a unifying epistemological orientation to the analysis of vital phenomena, and that the structure of John Hunter's physiological problematic should be regarded as an instance of this unity.

158. I have restricted my argument to the disposition of parts, and to what Hunter terms "local diseases." In speaking of diseases of the whole "constitution," he ascribes dispositions to the constitution itself; in these cases disposition refers to a general condition of the body. Dispositions, therefore, like properties that may be ascribed to whole organs, are hierarchically structured.

159. *Lectures,* p. 301. I shall return to this important notion below.

160. Ibid., p. 301–2, footnote.

161. Ibid., p. 301.

162. *Treatise,* pp. 3–5; likewise, the second subsection of Hunter's *Treatise on the Venereal Disease,* in *Works,* 2: 132–33, is entitled "Of Morbid Actions Being Incompatible with Each Other."

163. *Treatise,* pp. 3–4.

164. Ibid., p. 4.

165. *Observations,* p. 96.

166. *Treatise,* p. 239. It is in explaining these processes substantively, rather than conceptualizing disease dispositions abstractly, that Hunter's physiological theory affords

practical application to pathology; hence the logic of the *Treatise on the Blood, Inflammation, and Gun-Shot Wounds,* which discusses the physiological interactions between the solids and the fluids, and their role in the healing of lesions and wounds. By defining inflammation as a restorative process, Hunter may be said to place himself in the tradition of Sydenham and Stahl: L. J. Rather, *Addison and the White Corpuscles: An Aspect of Nineteenth Century Biology* (Berkeley and Los Angeles: University of California Press, 1972), pp. 5–6, 146, 148.

167. *Treatise,* p. 398. See *Lectures,* pp. 296–97 for comments on habituation to effluvia.

168. Cf. the better-known and explicitly formulated theory of Xavier Bichat, *Recherches physiologiques,* pp. 80–84. While Hunter conceptualized the ontological basis of vitality as the replacement of one kind of property by another, Bichat argued that the variability of vital forces results from a struggle between physical and organic laws. See Everett Mendelsohn, "Physical Models and Physiological Concepts: Explanation in Nineteenth Century Biology, *British Journal for the History of Science* 2 (1965): 214–15; Albury, "Experiment and Explanation," pp. 63–67.

169. *Lectures,* p. 220.

170. *Treatise,* p. 365. I have not had a chance to consult a recent monograph which discusses Hunter's "tissue pathology": Othmar Keel, *La Généalogie de l'histopathologie, une révision déchirante* (Paris: J. Vrin, 1979).

171. Cf. Bichat's distinction between the animal or external life, and the organic or internal life of an organism: Xavier Bichat, *Recherches physiologiques,* pp. 1–162. See A. Arène, "Essai sur la philosophie de Xavier Bichat," *Archives d'anthropologie criminelle, de médecine légale, et de psychologie, normal et pathologique* (1911): 753–825.

172. *Lectures,* p. 243, and quotations in paragraph following.

173. Ibid., p. 245.

174. See note 82 above. Hunter makes this distinction with particular reference to whales, whose internal oeconomy is analogous to that of the "terrestrial quadrupeds," but whose external form "fits them for dividing the water in progressive motion." The physiologist, Hunter laments quaintly, is "too often obliged to reason from analogy where information fails, which must probably ever continue to be the case, from our unfitness to pursue our researches in the unfathomable waters. This unfitness does not arise from that part of our oeconomy on which life and its functions depend, for the tribe of animals which is to be the subject of this Paper has, in that respect, the same oeconomy as man, but from a difference in the mechanism by which our progressive motion is produced." *Observations,* "Observations on the Structure and Oeconomy of Whales," p. 331.

175. *Observations,* p. 214.

176. *Lectures,* p. 260.

177. Ibid., p. 264.

178. *Essays,* 1: 152.

179. Quotations in this paragraph are from *Observations,* pp. 209–10.

180. *Lectures,* p. 268. Cf. *Essays,* 1: 164.

181. Owsei Temkin, "The Classical Roots of Glisson's Doctrine of Irritability," *Bulletin of the History of Medicine* 38 (1964): 297–328; Charles Daremberg, *Histoire des sciences médicales* (Paris: J.-B. Baillière, 1870), 2: 650–73; E. Bastholm, *The History of Muscle Physiology,* pp. 216–39. The Haller–Whytt dispute is discussed in detail in R. K. French, *Robert Whytt, the Soul, and Medicine* (London: The Wellcome Institute of the History of Medicine, 1969), pp. 63–76. On the Montpellier school see note 149; Théophile Bordeu's opposition to the Hallerian concept of sensibility is discussed in Herbert

Dieckmann, "Théophile Bordeu und Diderots *Rêve de D'Alembert*," *Romanische Forschungen* 52 (1938): 81–82. Denis Diderot, *Éléments de physiologie*, ed. and intro. Jean Mayer (Paris: Marcel Didier, 1964), compiled between 1774 and 1780, may be treated as a précis of the rise of sensibility. A general survey of theories of irritability and sensibility in this period is Thomas S. Hall, *History of General Physiology*, 2: 66–106.

182. An excellent analysis of the concept of the *sensorium commune* is to be found in Figlio, "Theories of Perception and the Physiology of Mind."

183. *Treatise*, p. 116.

184. *Lectures*, p. 317. John Abernethy, believing himself faithful to Hunter's opinion, followed an eighteenth-century tradition in his Royal College of Surgeons lectures in identifying the vital material with one of the "imponderable fluids"–electricity. His doctrine was attacked by William Lawrence in subsequent Hunterian lectures, who, correctly it would seem, stressed Hunter's belief in the physiological uniqueness and ineffability of vital properties. John Abernethy, *Introductory Lectures Exhibiting Some of Mr. Hunter's Opinions Respecting Life and Diseases, Delivered Before the Royal College of Surgeons London, in 1814 and 1815* (London: Longman, Hurst, Rees, Orme, and Brown, 1823); William Lawrence, *An Introduction to Comparative Anatomy and Physiology; Being the Two Introductory Lectures Delivered at the Royal College of Surgeons . . . 1816* (London: R. Carlile, 1823), pp. 71–78. For discussions of the Abernethy/Lawrence debate, see June Goodfield-Toulmin, "Some Aspects of English Physiology"; Owsei Temkin, "Basic Science, Medicine, and the Romantic Era," *Bulletin of the History of Medicine* 37 (1963): 97–129; Figlio, "The Metaphor of Organization." The hypothesis that vital phenomena might be caused by micro-forces of an electrical nature had been entertained, privately, by Isaac Newton in the first decade of the eighteenth century. See J. E. McGuire, "Force, Active Powers," pp. 177–187. Eighteenth-century conceptions of the vital principle as an electrical substance, and John Hunter's relation to them, are discussed in Philip C. Ritterbush, *Overtures to Biology*, pp. 1–56, 186–97. See also Robert Schofield, *Mechanism and Materialism*, pp. 208–9; Roderick W. Home, "Electricity and the Nervous Fluid," *Journal of the History of Biology* 3 (1970): 235–51.

185. Most notably the "physiological medicine" of François J. V. Broussais (1772–1838), whose *Examen de la doctrine médicale généralement adoptée et des systèmes modernes de nosologie* (Paris: Gabon, 1816) precipitated the downfall of "ontological" nosology in Paris. Like Hunter, Broussais regarded local inflammatory lesions, caused by irritation, as evidence of pathological malfunction. See Erwin H. Ackerknecht, *Medicine at the Paris Hospital 1794-1848* (Baltimore: Johns Hopkins Press, 1967), pp. 61–80. Cf. Alain Rousseau, "Gaspard-Laurent Bayle (1774-1816). Le théoricien de l'École de Paris," *Clio Medica* 6 (1971): 205–11.

186. See note 103.

187. Cf. Erasmus Darwin, *Zoonomia*, 2, "Preface," p. i.: "All diseases originate in the exuberance, deficiency, or retrograde action of the faculties of the sensorium, as their proximate cause; and consist in the disordered motions of the fibres of the body, as the proximate effect of the exertions of those disordered faculties."

188. This is not to deny that partisan debates over the relative value of theoretically sanctioned therapeutics were highly elaborate. For examples of the genre, see William Yates and Charles Maclean, *A View of the Science of Life; on the Principles Established in the Elements of Medicine of the Late Celebrated John Brown, M.D.* (Calcutta: Thomson and Ferris, 1797); Henrique Xavier Baeta, *Comparative View of the Theories and Practice of Drs. Cullen, Brown and Darwin in the Treatment of Fevers and of Acute*

Rheumatism (London: J. Johnson, 1800). The tremendous influence of Brunonian medicine in Italy and Germany is discussed in John Thomson, *An Account of . . . William Cullen*, 2: 354–506; Guenter B. Risse, "The History of John Brown's Medical System in Germany during the Years 1790–1806," Ph.D. dissertation, University of Chicago, 1971. See comments on the relation between academic medicine and traditional therapeutics in Charles E. Rosenberg, "The Therapeutic Revolution."

189. *Treatise*, p. 333.

190. See note 181. The most detailed study of the origins of neural pathology that I have seen is Inci Altug Bowman, "William Cullen (1710–90) and the Primacy of the Nervous System." Bowman suggests that the differences between Hunter and Cullen should be viewed in terms of a wider cardiocentric/cephalocentric controversy in the eighteenth century, echoing disputes between the followers of Aristotle and those of Galen (p. 112). I have tried to show, however, that while Hunter certainly repudiated cephalocentric pathology, his theory of disease was essentially built upon solidist conceptions. Christopher Lawrence is surely correct to suggest that a study of a physician such as George Fordyce, who began his career as a student of Cullen, but developed it as John Hunter's close intellectual companion, would shed much light on these differing physiological conceptions and their roots. Christopher Lawrence, "The Nervous System and Society," p. 35. Yet Lawrence's sociological explanations are tendered without reference to epistemic continuities between the vitalisms of Edinburgh, London, and the Continent.

191. See Georges Canguilhem, "John Brown (1735–1788). La théorie de l'incitabilité de l'organisme et son importance historique," *Proceedings of the 13th International Congress of the History of Science* (Moscow: Nauka, 1974), 9: 141–46. Cf. Guenter B. Risse, "The Quest for Certainty in Medicine: John Brown's System of Medicine in France," *Bulletin of the History of Medicine* 45 (1971): 1–12; idem, "Kant, Schelling, and the Early Search for a Physiological 'Science' of Medicine in Germany," *Journal of the History of Medicine* 27 (1972): 145–58.

192. John Edmonds Stock, *Memoirs of the Life of Thomas Beddoes, M.D.* (London: John Murray, 1811), p. 132.

193. William Lawrence, *Lectures on Physiology*, p. 87. See note 182.

194. *Treatise*, p. 429; *Lectures*, p. 248; *Observations*, p. 167.

195. *Lectures*, p. 249.

196. Ibid., p. 247.

197. Ibid., p. 247–48.

198. Ibid., p. 264. Hunter's distinction between the sensible and living principles leads to similar conclusions to those reached by Théophile Bordeu in his distinction between *sentiment* and *sensibilité*. In each case the former concept refers to the provenance of sentience and the head, the latter to the internal animal oeconomy and an organic sensibility centered in the region of the stomach. Bordeu's doctrine of the epigastric center no doubt owed a debt to the seventeenth-century scientist and physician Jean-Baptiste van Helmont (1577–1644), who located the controlling *archaeus*, or immanent vital spirit, in the stomach. See Jacques Roger, *Les Sciences de la vie*, pp. 629–30; Walter Pagel, "The Religious and Philosophical Aspects of van Helmont's Science and Medicine," *Bulletin of the History of Medicine*, supplement no. 2 (1944): 36–38.

199. *Lectures*, p. 317. Typical of Hunter's arguments for the lack of importance of the nerves in the oeconomy is the following: "I have observed, that in inflammation the vessels become larger, more blood passes, and there appear to be more actions taking place; but the nerves do not seem to undergo any change. The nerves of the gravid uterus are the same as when it is in the natural state; neither do the branches of the fifth and

seventh pair of nerves in the stag become larger [when his horns are growing, though the external carotids increase in size] ; and in inflammation of the nerves their blood-vessels are enlarged, and have coagulating lymph thrown into their interstices, but the nerve itself is not increased, so as to bring the part to the state of a natural part, fitted for acute sensation, which shows that the motions of the nerves have nothing to do with the oeconomy of the part; they are only the messengers of intelligence and orders." *Treatise*, p. 333.

200. *Lectures*, p. 236.

201. *Treatise*, p. 112.

202. *Lectures*, p. 229.

203. *Essays*, 1: 113. Cf. *Lectures*, p. 231. Hunter probably had in mind a process analogous to dephlogistication, as suggested in Thomas S. Hall, *History of General Physiology*, 2: 111. In the *Treatise* Hunter discusses respiration simply from the viewpoint of unexplained changes in the color of the blood. He regarded the red globules as the least important part of the blood, since they play no evident role in the major function of coagulation: ". . . we must conclude them not to be the important part of the blood in contributing to growth, repair, etc. Their use would seem to be connected with strength, for the stronger the animal the more it has of the red globules." *Treatise*, p. 68. For a brief survey of eighteenth-century theories of the constituent parts of the blood and their functions, see George Gulliver, ed. and intro., *The Works of William Hewson, F.R.S.* (London: The Sydenham Society, 1846), "Introduction on the Life and Writings of Hewson, with a History of the Coagulation of the Blood," pp. xxv–xlviii.

204. Hunter's principal role in the humoralist revival of the late eighteenth century, and in the wave of interest in tissue formation and the microvascular processes of inflammation, which continued into the middle of the nineteenth century, is discussed in L. J. Rather, *The Genesis of Cancer: A Study in the History of Ideas* (Baltimore: Johns Hopkins University Press, 1978), pp. 41–45; idem, *Addison and the White Corpuscles*, pp. 6–8, 217–18; John V. Pickstone, "Globules and Coagula: Concepts of Tissue Formation in the Early Nineteenth Century," *Journal of the History of Medicine* 28 (1973): 336–56. John Hunter's "coagulable lymph" would seem to be analogous to the "cystoblastema" of Theodor Schwann (1810–1882), the formless primitive sap out of which cells were held to crystallize: L. J. Rather, "Some Relations Between Eighteenth-Century Fiber Theory," pp. 198–200; Russell C. Maulitz, "Schwann's Way: Cells and Crystals," *Journal of the History of Medicine* 26 (1971): 422–37.

205. *Treatise*, p. 460.

206. Everard Home, "Experiments and Observations on the Growth of Bones, From the Papers of the Late Mr. Hunter," *Observations*, p. 318.

207. Thus Hunter's theory of absorption distanced itself explicitly from theories which attributed the destruction of organic parts to mechanical stress and friction; e.g. Albertus Haller, *First Lines of Physiology*, ed. William Cullen (Edinburgh: Charles Elliot, 1786; facs. reprint New York: Johnson Reprint Corporation, 1966), 2: 241–43 (sections 957–61).

208. William Hunter, *Two Introductory Lectures*, p. 58, original emphasis.

209. William Hunter's claim to priority of discovery was disputed by Alexander Monro *secundus;* see Nellie B. Eales, "The History of the Lymphatic System, with Special Reference to the Hunter-Monro Controversy," *Journal of the History of Medicine* 29 (1974): 280–94. A useful account of the Hunter school's work on the lymphatics is William Hewson, *Experimental Inquiries. Part II: A Description of the Lymphatic System* (1774), in *The Works of William Hewson*, pp. 113–204. John Hunter's experiments are recorded in William Hunter, *Medical Commentaries, Part I*, pp. 38–53, of

which pp. 42-49 cite a John Hunter manuscript; reprinted in *Observations,* pp. 299-307. See also *Lectures,* pp. 250-58; *Treatise,* pp. 459-87. The late eighteenth-century doctrine of the lymphatics and its fate is put into context and evaluated historiographically in John V. Pickstone, "The Origins of General Physiology in France with Special Emphasis on the Work of R. J. H. Dutrochet," Ph.D. dissertation, University of London, 1973, pp. 125-61. Magendie's experimental investigations of absorption are described in detail in John E. Lesch, "The Origins of Experimental Physiology and Pharmacology in France, 1790-1820: Bichat and Magendie," Ph.D. dissertation, Princeton University, 1977, pp. 209-85. See also Richard Owen's commentary, *Observations,* pp. 307-14.

210. *Treatise,* p. 195.

211. In the extremely influential iatromechanical physiology of Herman Boerhaave (1668-1738) of Leiden, vessels are the simplest structural elements of the body: G. A. Lindeboom, "Boerhaave's Concept of the Basic Structure of the Body," *Clio Medica* 5 (1970): 203-8; Lester S. King, "Introduction," Albertus Haller, *First Lines of Physiology,* pp. xxiii-xxxii.

212. *Treatise,* p. 459.

213. *Observations,* p. 213.

214. *Lectures,* p. 264.

215. *Treatise,* p. 147.

216. I have borrowed this distinction between theoretical and classificatory discourses from Keith Tribe, *Land, Labour and Economic Discourse* (London: Routledge and Kegan Paul, 1978). It implies no Whiggism, or the notion of a teleological emergence of science from an "ideological" precursor, but serves, on the contrary, to express the discontinuity between a structure of knowledge in which "discourse" is present, and a structure in which it is absent. On this Foucauldian notion, see discussion below, and Foucault, *The Order of Things,* pp. 303-5; Devereaux Kennedy, "Michel Foucault," pp. 278-80. In the concluding section of this essay I develop parallels between Tribe's analysis of eighteenth-century political oeconomy and my own explication of animal oeconomy. An interpretation of the Hunterian doctrine of the lymphatics similar to the one above is made by John V. Pickstone in his doctoral thesis "The Origins of General Physiology in France." Pickstone uses the eighteenth-century investigation of the absorption system, an example of the problematic which posed the question of the functions of specific organs, to illustrate the emergence with experimentalists such as Magendie of a new problematic in the early nineteenth century, one constitutive of a general physiology which posed the question of the mechanisms of general vital actions. See also J. V. Pickstone, "Absorption and Osmosis: French Physiology and Physics in the Early Nineteenth Century," *The Physiologist* 20 (1977): 30-37. Cf. Joseph Schiller, *Claude Bernard et les problèmes scientifiques de son temps* (Paris: Éditions du Cèdre, 1967), pp. 43-58.

217. William Harvey, *On Animal Generation,* in *The Works of William Harvey, M.D.,* trans. and ed. Robert Willis (London: The Syndenham Society, 1847; facs. reprint New York: Johnson Reprint Corporation, 1965), p. 381.

218. Walter Pagel, "Harvey and Glisson on Irritability: With a Note on Van Helmont," *Bulletin of the History of Medicine* 41 (1967): 497-514; idem, *William Harvey's Biological Ideas: Selected Aspects and Historical Background* (New York: Hafner, 1967), pp. 251-65. On the seventeenth-century doctrine of immanence as a monistic view of living matter, see also idem, "The Reaction to Aristotelianism in Seventeenth Century Biological Thought," in *Science, Medicine and History,* ed. E. Ashworth Underwood (London: Oxford University Press, 1953), 1: 500-9. See note 181, and Theodore M. Brown, "Physiology and the Mechanical Philosophy in Mid-Seventeenth Century England," *Bulletin of the History of Medicine* 51 (1977): 38-44, 54.

219. *Treatise*, p. 34.

220. Ibid., p. 33.

221. Cf. Albury, "Experiment and Explanation," p. 84: "The theory of vital properties was an unchallenged feature of physiology from its introduction by Glisson until
the attack mounted against it in the first decade of the nineteenth century. It underlay
the great physiological controversies for more than a hundred years and defined the
theoretical boundaries within which they were disputed. Mechanists, vitalists, and
animists offered competing explanations of the vital properties; and when Haller and
the Montpellier physicians debated methodology, the argument centered on how best
to study these properties."

222. *Essays*, 1: 18.

223. Ibid., 1: 60.

224. *Observations*, p. 243. Cf. Théophile Bordeu, *Recherches anatomiques*, p. 187,
translated and quoted by Albury, "Experiment and Explanation," p. 90. "Thus the least
part may be regarded as forming, so to speak, a separate body. . . . In order to appreciate
properly the particular action of each part, let us compare the living body to a swarm
of bees which gathers into little clusters and which hangs from a tree like a bunch of
grapes. The saying of a celebrated Ancient, that one of the viscera of the lower abdomen
was an animal within an animal, has met with approval: each part is, so to speak, undoubtedly not an animal but rather a kind of separate machine which contributes in its
fashion to the general life of the body." Sergio Moravia, "From *homme machine* to
homme sensible," pp. 54–60, provides a useful discussion of Bordeu's model of the
animal body as a "federation of organs," but does so from the viewpoint of the vitalist-
mechanist debate, and modern theories of the peripheral nervous system, rather than
from the viewpoint of the problematic of organic relations developed here. From the
latter viewpoint, compare Bordeu's metaphor with an early use of the word "organism"
in Leibniz: "j'ay dit non pas absolument, que l'organisme est essentiel à la matiere,
mais à la matiere arrangée par une sagesse souveraine. Et c'est pour cela aussi que je
definis l'Organisme, ou la Machine naturelle, que c'est une machine dont chaque partie
est machine, et par consequent que la subtilité de son artifice va à l'infini, rien n'estant
assez petit pour estre negligé, au lieu que les parties de nos machines artificielles ne sont
point des machines. C'est là la difference essentielle de la Nature et de l'Art, que nos
modernes n'avoient pas assez consideréé." *Die philosophischen Schriften von Gottfried
Wilhelm Leibniz*, ed. C. I. Gerhardt (Berlin: Weidmannsche, 1887), 3: 356, letter from
Leibniz to Lady Masham, 1704. I am grateful to Camille Limoges for refering this passage to me. See note 140 on notions of the animal machine as a multiplicity of independent powers; i.e., as an "organized body" in the eighteenth-century sense; and
historiographic discussion, note 144.

225. *Lectures*, pp. 272-3.

226. Ibid., p. 318.

227. Ibid.

228. *Lectures*, p. 327.

229. Hunter attributed the phenomena of plant motion—diurnal movements, for
example—to the inherent properties of "vegetable life." Hence, he termed plants capable
of motion "sensitive plants," and suggested that their power of action is analogous to
the irritability of animals. Some plants, such as *Mimosa pudica*, also known as "the
Sensitive-plant," have special structures fitted for motion, Hunter believed. See *Essays*,
1: 341, 356-57; *Observations*, pp. 196-207; and note 71. Significantly the concept of
plant sensitivity appeared at the same time as the concept of animal irritability, at the
beginning of the classical period: Charles Webster, "The Recognition of Plant Sensitivity

by English Botanists in the Seventeenth Century," *Isis* 57 (1966): 5–23; Philip C. Ritterbush, *Overtures to Biology*, pp. 141–57.

230. *Observations*, p. 249.

231. *Lectures*, p. 326. Concepts of sympathy and its mechanisms at the time of Hunter are discussed briefly in C. Barker Jørgensen, *John Hunter, A. A. Berthold*, pp. 7–9, 22–24. Jørgensen correctly stresses that Hunter performed transplantation experiments to elucidate not the general physiological processes that were to constitute the objects of the science of endocrinology, but the relative powers of different parts and organs, and the specific sympathetic relations between them. See note 120. Hunter again adopts a view here different from those who stressed the basic mediational role of the nerves in vital action. Robert Whytt, for example, maintained that all sympathy presupposes feeling, and is therefore effected by the intervention of the soul via the nerves. But like Hunter he believed that sympathy constitutes the basic mechanism of organic coordination in health. See R. K. French, *Robert Whytt*, pp. 31–38. The medical investigation of sympathy in Hunter had a sequel above all in John Abernethy, *Surgical Observations on the Constitutional Origin and Treatment of Local Diseases* (London: Longman, Hurst, Rees, and Orme, 1809).

232. *Lectures*, p. 248. Cf. *Essays*, 1: 37: "Perhaps the declension of animals from the most perfect to the most imperfect is in a regular order or progression; but that progress is far from being an equal one; for the differences or distances between, or amongst, the most perfect are great and obvious; but when we come among the imperfect they are much closer, or less observable, if at all. This makes the most perfect but few in number in comparison with the others; and, in these, increases the number of genera and species. . . ."

233. *Essays*, 1: 36.

234. *The Natural History of the Human Teeth*, in *Works*, 2: 55–56.

235. Cf. *Essays*, 1: 19.

236. *A Practical Treatise on the Diseases of the Teeth*, in *Works*, 2, "The Natural History of the Human Teeth, Part II," pp. 106–7.

237. *Essays*, 1: 203. Hunter goes on to argue that such comparison is not, in fact, possible, because the earliest stages of formation in embryogenesis are invisible to the naked eye, and hence as I explain below, beyond the epistemological limits of anatomical knowledge. Cf. *Observations*, pp. 268–69.

238. E.g. Joseph Needham, *A History of Embryology* (Cambridge: Cambridge University Press, 1934), p. 203. On the early history of organic-historical theories of recapitulation, see Owsei Temkin, "German Concepts of Ontogeny and History around 1800," *Bulletin of the History of Medicine* 24 (1950): 227–46; William Coleman, "Limits of the Recapitulation Theory: Carl Friedrich Kielmeyer's Critique of the Presumed Parallelism of Earth History, Ontogeny, and the Present Order of Organisms," *Isis* 64 (1973): 341–50.

239. William Albury, and Michael Gross in his doctoral dissertation "Function and Structure in Nineteenth Century French Physiology," have argued that Georges Cuvier and François Magendie shared a conception of general metabolic processes as the defining characteristic of life, and Albury has demonstrated that a common epistemological reorganization made possible both Cuvier's conception of internal anatomical relations as organismic conditions of existence, and Magendie's conception of organic functions such as nutrition as not the properties of parts, but the effects of the internal organization of organs. Albury, "Experiment and Explanation," pp. 87–94. Cf. Michael Gross, "The Lessened Locus of Feelings"; Frederic L. Holmes, "The Transformation of the Science of Nutrition," *Journal of the History of Biology*, 8 (1975): 135–44, especially

139–40; Franklin C. Bing, "The History of the Word 'Metabolism,'" *Journal of the History of Medicine* 26 (1971): 158–80, especially 172–73. Drawing upon the work of Michel Foucault in a similar manner to Albury, Bernard Balan has suggested that *within* the classical period, and in tension, were two distinct approaches to the animal oeconomy, the first focusing upon anatomical organization and the properties of parts, the second upon the intimate processes and functions of life. This second problematic, Balan argues, was developed by Cuvier in his break with the naturalists, transforming the "oeconomy" into an "economy." However, Balan mistakenly interprets John Hunter's concept of the bricklayers as a general physiological theory of chemical-tissual exchanges, similar to Cuvier's non-analytic reduction of animal functions to the transformations of fluids. B. Balan, "Premières recherches sur l'origine et la formation du concept d'économie animale," *Revue d'histoire des sciences* 28 (1975): 291–5, 325–6. I have not seen his *L'Ordre et le temps: l'anatomie comparée et l'histoire des vivants au XIXe siècle* (Paris: J. Vrin, 1979).

240. Foucault, *The Order of Things*, pp. 46–77.

241. Albury, "Experiment and Explanation," p. 85. Frederic Holmes has argued that "extension of Foucault's doctrine" to areas of biology other than taxonomy is untenable; and indeed Foucault has stressed that his object of study was the "law of construction" of methods of classification. Frederic L. Holmes, "Review of François Jacob, *La logique du vivant,*" *Studies in History of Biology* 1 (1977): 218; Foucault, "La situation de Cuvier," p. 71. The purpose of this study of John Hunter, however, has not been to apply, or test, a pregiven doctrine, but to utilize a similar methodology, with the shared aim of uncovering the "law of construction" of a specific theory of the animal oeconomy. Thus, while Holmes is obviously correct to point out that physiologists from at least the time of Galen had been interested in the interconnections of functional systems, which Foucault shows were ignored by naturalists in the classical period, one need not deny that a common epistemology and discourse may provide continuity at a hidden level between these areas of biology—as Albury has claimed—and one cannot conversely presume that a continuity exists within discourse upon physiological functions from classical antiquity to Cuvier. Cf. W. R. Albury and D. R. Oldroyd, "From Renaissance Mineral Studies to Historical Geology, in the Light of Michel Foucault's *The Order of Things,*" *British Journal for the History of Science* 10 (1977): 187–215.

242. On the importance of Condillac's logic for the *Idéologues,* see Albury, "The Logic of Condillac and the Structure of French Chemical and Biological Theory"; A. Arène, "Essai sur la philosophie de Xavier Bichat"; Owsei Temkin, "The Philosophical Background of Magendie's Physiology"; George Rosen, "The Philosophy of Ideology and the Emergence of Modern Medicine in France," *Bulletin of the History of Medicine* 20 (1946): 328–39; M. Laignel-Lavastine, "Sources, principes, sillage et critique de l'oeuvre de Bichat," *Bulletin de la société francaise de philosophie* 46 (1952): 4–28. W. Riese, "La méthode analytique de Condillac et ses rapports avec l'oeuvre de Phillippe Pinel," *Revue philosophique* 158 (1968): 321–36. Georges Gusdorf, *Dieu, la nature, l'homme,* pp. 481–5; Sergio Moravia, "Philosophie et médecine en France"; Robert Marc Freidman, "The Creation of a New Science: Joseph Fourier's Analytical Theory of Heat," *Historical Studies in the Physical Sciences* 8 (1977): 73–99. Albury has stressed in his dissertation (e.g. pp. 90–91) that Condillac's philosophy should not be assigned ultimate responsibility for the acceptance of analysis and "sensualism" by the *Idéologues,* but should be seen as giving form to their response to pre-existing needs. With this perspective one can comprehend why, in general terms, as Lester King observes, "Haller acted as if he were putting into practice the doctrines that Condillac described and advocated." Lester S. King, "Introduction," Albert Haller, *First Lines of Physiology,* p. lvi.

243. *Essays*, 1: 11.

244. Drewry Ottley, "The Life of John Hunter," p. 56. In the original manuscript: "but why think, why not trie the Expt." *Letters from the Past: From John Hunter to Edward Jenner,* ed. A. J. Harding Rains (London: Royal College of Surgeons of England, 1976), p. 9. See note 23.

245. *Treatise*, pp. 93-94.

246. Ibid., pp. 182.

247. Ibid., p. 184.

248. Ibid., p. 448.

249. Cf. Foucault, *The Birth of the Clinic*, pp. 167; W. R. Albury and D. R. Oldroyd, "From Renaissance Mineral Studies to Historical Geology," pp. 191–201.

250. *Treatise,* p. 60, footnote.

251. Ibid., p. 14. Cf. *Essays,* 1: 204: "[Magnifying] glasses lead us back far beyond what the naked eye reaches; but these only show us the order of priority in the formation of parts. However, human wisdom can go no further than into the distinction of parts, with their actions and uses when formed." The microscope slide collection in Hunter's possession had been bought from the sale of William Hewson's museum; see Jessie Dobson, "John Hunter's Microscope Slides," *Annals of the Royal College of Surgeons of England* 28 (1961): 175–88. The epistemological basis of the rejection of microscopy in life science from Sydenham and Locke at the beginning of the classical period to the "cult of observation in the Paris Clinical School" at its end is discussed in David E. Wolfe, "Sydenham and Locke on the Limits of Anatomy," *Bulletin of the History of Medicine* 35 (1961): 193–220.

252. *Observations,* p. 82.

253. Cf. ibid., pp. 81–121; *Essays,* 1: 146–48; George Fordyce, *A Treatise on the Digestion of Food* (London: J. Johnson, 1791), especially pp. 153–56.

254. *Observations,* p. 82. For the Condillacian *Idéologues* a true science was a "well-made language." See H. B. Acton, "The Philosophy of Language in Revolutionary France," *Proceedings of the British Academy* 45 (1959): 199–219.

255. Foucault, *The Order of Things,* p. 232.

256. Idem, *The Birth of the Clinic,* p. 129.

257. Figlio, "The Metaphor of Organization." It will be apparent that this epistemological issue has a political content. For, if ideological investment of scientific knowledge occurs at the level of an underlying problematic or set of root metaphors, the critiques of science as "interested" knowledge which attempt only to identify "abuses" of science, to separate "good science" from "bad science," or to purge science of ideology, run the risk of immunizing from political examination the most basic compounding of language and social purposes in the very structure of a scientific discourse as a whole. For example, as Donna Haraway argues, much of the current argument against sociobiology is vitiated by its refusal to recognize sociobiology as anything other than the latest episode in a continuous tradition of ideological perversions of biology for explicitly racist, sexist, and classist ends. This approach fails to identify the genuine and novel power of the "new synthesis" as ideology, since this resides in its *biological* power as a model of the cybernetic and communications approach which characterizes the leading edges of contemporary biology. An adequate comprehension of the ideological nature of sociobiology, Haraway argues, necessitates critique of contemporary biology as a total discourse integral to the managerial and control practices of late capitalism. Donna Haraway, "The Biological Enterprise: Sex, Mind, and Profit from Human Engineering to Sociobiology," *Radical History Review* 20 (Spring/Summer 1979): 206–37.

258. I refer in the following paragraphs primarily to the political oeconomists and

conjectural historians of the Scottish Enlightenment. See Gladys Bryson, *Man and Society: The Scottish Enquiry of the Eighteenth Century* (Princeton: University Press, 1945); J. W. Burrows, *Evolution and Society* (London: Cambridge University Press, 1966), pp. 1–64.

259. *Lectures*, p. 273.

260. Ibid., p. 220.

261. *Essays*, 1: 119; emphasis added.

262. Keith Tribe, *Land, Labour and Economic Discourse.*

263. Ibid., p. 145.

264. Cf. for example, Peter Mark Roget, *Animal and Vegetable Physiology Considered with Reference to Natural Theology,* Bridgewater Treatise V (London: William Pickering, 1840), 1: 1–49; 2: 1–13, 62–82.

265. Cf. Michel Foucault, *The History of Sexuality. Volume I: An Introduction,* trans. Robert Hurley (New York: Pantheon Books, 1978); Robert Young, "The Historiographic and Ideological Contexts of the Nineteenth-Century Debate on Man's Place in Nature," in *Changing Perspectives in the History of Science: Essays in Honour of Joseph Needham,* ed. Mikuláš Teich and Robert Young (Dordrecht: D. Reidel, 1973), pp. 350, 376, 384; idem, "Functionalism," mimeograph, 1971; Roger Cooter, "The Power of the Body: The Early Nineteenth Century," in Barry Barnes and Steven Shapin, eds., *Natural Order,* pp. 73–92; Barbara Haines, "The Interrelations Between Social, Biological, and Medical Thought, 1750–1850: Saint-Simon and Comte," *British Journal for the History of Science* 37 (1978): 19–35; Keith Michael Baker, *Condorcet: From Natural Philosophy to Social Mathematics* (Chicago: University of Chicago Press, 1975), pp. 371–82; Frank E. Manuel, "From Equality to Organicism," *Journal of the History of Ideas* 17 (1956): 54–69; Robert A. Nisbet, "The French Revolution and the Rise of Sociology in France," *American Journal of Sociology* 49 (1943): 156–64; Judith E. Schlanger, *Les Metaphores de l'organisme* (Paris: J. Vrin, 1971). The conditions of possibility of a biological conception of life are of course subject to historical change, and perhaps no longer obtain. See Michel Foucault, *The Order of Things,* pp. 340–43, 386–87; François Jacob, *The Logic of Life,* pp. 299–324; Edward J. Yoxen, "The Social Impact of Molecular Biology," Ph.D. dissertation, University of Cambridge, 1977; idem, "Where Does Schröedinger's 'What Is Life?' Belong in the History of Molecular Biology?" *History of Science* 17 (1979): 17–52, especially 36–43; Donna Haraway, "The Biological Enterprise"; idem, "Signs of Dominance: From a Physiology to a Cybernetics of Primate Society, C. R. Carpenter, 1930–70," mimeograph, 1979.

The Göttingen School and the Development of Transcendental Naturphilosophie in the Romantic Era

Timothy Lenoir*

A problem, tantalizing in its possible implications, that has persistently thwarted the efforts of historians is the relationship between empirical science and the speculative movement in philosophy and literature at the beginning of the nineteenth century, known as *Naturphilosophie*. While some scholars have regarded Naturphilosophie as a skeleton in the closet of nineteenth-century science,[1] others have indicated that it may have had a positive influence on several major discoveries.[2] There have been severe difficulties in interpreting the substantive contribution of Naturphilosophie to the development of science, however. One central difficulty in explaining how naturphilosophic systems were able to reign supreme in the German scientific community from 1800 to 1830 lies, of course, in deciphering the actual scientific content of the philosophies of nature proposed by the likes of Schelling, Oken, Hegel, and Carus, and the extent to which they incorporated a careful consideration of the contemporary scientific literature. The verdict on this issue has by no means been unambiguous: Some investigators have argued that in their disdain for empirical research the Naturphilosophen

I owe a special debt of gratitude to Professor William Coleman for extended discussion of the problems treated in this paper and to Professor Reinhard Löw of the Ludwig-Maximilians-Universität München for many hours of patient discussion of Kant's theory of teleology and its significance for the life sciences in the Romantic era. Research support from the National Science Foundation is also gratefully acknowledged.
*Department of History, University of Arizona.

were attempting to return science to a simpler age.[3] Others have argued that the heart and soul of Naturphilosophie lay in empirical research. Those who defend this latter interpretation—an interpretation that is in rapid ascendance in the literature—point out that while Romantics such as Novalis, who had been trained in the sciences at the Bergakademie in Frieberg under Werner, demonstrated a strong scientific bent, other Naturphilosophen such as Goethe, Ritter, Oken, and Carus conducted extensive empirical researches themselves.[4] The potential sources of confusion in assessing this issue emerge clearly in the work of Hegel; for while he defended a conception of matter based on the four elements, earth, air, fire, and water, it is clear that he was deeply immersed in the chemical literature of the day and that he understood it well.[5]

Another problem in assessing the relationship of Naturphilosophie to science is rooted in the fact that no single system of natural philosophy is characteristic of the entire Romantic period. From its first appearance and throughout its stormy career, for instance, the Naturphilosophie of Schelling and his school was severely criticized.[6] When we turn to the writings of these critics, however, we discover many of the same conceptual elements and almost invariably refer to the same empirical data.[7] Concern is quite naturally generated about identifying the real substance of the issues being debated. Rather than a single systematic approach to nature, it seems more appropriate to regard the science of this era as having been formed from a common fund of scientific concepts and methods, metaphysical predispositions and episte-mological concerns which received differing emphases in the various approaches to natural philosophy of the period. In order to assess the bases for these different styles of Naturphilosophie consideration will have to be given to the role not only of substantive philosophical and scientific issues but of personal factors as well. But a full understanding of these complex issues may ultimately await the exploration of broadly based trends in the popular culture of the period as well as the roles of social and political movements in shaping preferences for organizing and interpreting this common fund of concepts.

A better understanding of this period has resulted from recent progress in dispelling the myth of a monolithic Romantic science, and in laying bare the outlines of different traditions of natural philosophy practiced in Germany between 1790 and 1830. This has been achieved chiefly through the efforts of Reinhard Löw, H. A. M. Snelders, and Dietrich von Englehardt. Von Englehardt has argued that three different traditions characterize the science of the Romantic era.

One tradition, which he identifies as Kantian, is *transcendental Naturphil-osophie*. In the spirit of Kant's critical writings this tradition views the role of philosophy as examining the logical and epistemological foundations of science by establishing the subjective contribution to experience, the a priori forms in terms of which empirical judgments are constituted, and the constraints on reason in constructing an interpretation of nature. The object of transcendental Naturphilosophie was not to explicate the proper method for abstracting lawlike generalizations from nature as given in experience. Rather, it aimed at "determining the *a priori* conditions for the possibility of experience, which is to provide the source from which general laws of nature are to be deduced."[8] Characteristic of this program is Kant's determination of the concept of matter in his *Metaphysische Anfangsgründe der Naturwissenschaft*. There, applying the categorical theory of his *Kritik der reinen Vernunft*, Kant argued that the concept of matter that must underlie mechanics cannot employ irreducible atoms but rather must invoke a dynamic interaction of attractive and repulsive forces emanating from nonmaterial points. This dynamic theory of matter, which had been proposed by Boscovich, became one of the central organizing concepts representative of the Kantian tradition of Naturphilosophie, and it was especially significant for the view of organic nature.

A second tradition of Naturphilosophie removed the boundaries of possible a priori knowledge of nature considered legitimate by transcendental Naturphilosophie. This second tradition is linked most closely with Schelling and is termed speculative or romantic Naturphilosophie by von Engelhardt. According to the speculative Romantics nature is a fundamental unity of matter, process, and spirit. The object of the philosophy of nature, according to this approach, is to construct the entire material system of nature from a single all-embracing unity, to establish the unfolding of the inorganic, organic, and finally the social and moral realms as the final objectification of potencies present in this original unity, which Schelling characterized alternately as the *Weltseele, Gott* or the *Absolut*. Characteristic of speculative thought is its claim that the dichotomy between empirical knowledge claims and the world of things in themselves crucial to Kantian or transcendental Naturphilosophie can be overcome in the act of "intellectual intuition," an empirical intuition in which the logical structure of appearances is also manifest. Also characteristic of this approach is its reliance upon polarity as the motive agent in the process of differentiating and objectifying the primitive unity at the basis of nature. Equally characteristic is the notion that the plant and animal kingdoms are each constituted from the metamorphosis of a fundamental unitary type, or *Urtyp,* and accordingly that organic nature can be perceived as a chain of beings. Perhaps most characteristic of speculative Naturphilosophie is the view that since nature is the manifestation of spirit, man must stand at the top of the chain of being.

Although it was not always clearly distinguished during the Romantic era, there was a third tradition of Naturphilosophie. This type, which von

Engelhardt calls metaphysical Naturphilosophie, was closely allied to the romantic or speculative tradition. Hegel, who was the main theoretician of this line of thought, in fact regarded the position developed by the young Schelling in his *Ideen zu einer Naturphilosophie* (1797) and in his *Von der Weltseele* (1798) as in fundamental agreement with the main lines of metaphysical Naturphilosophie; but there were certain tendencies in Schelling's thought that had been developed in an absurdly unphilosophical manner, with little knowledge of or concern for the empirical content of the sciences, by some of Schelling's most ardent followers, particularly Windischmann, Görres, and Steffens. In 1806 the differences in their outlooks led to a split between Hegel and Schelling. Principally, Hegel objected to the presence of mystical and irrational elements in Schelling's system, the so-called philosophy of identity. Moreover Schelling's attempt to deduce the material world completely from the self-activity of the Ego in terms of purely formal principles such as polarity, potential, and analogy, was objectionable in Hegel's view. Naturphilosophie could not possibly deduce the genesis of natural forms; its sole task consisted in bringing to the fore the logical structure of the system of nature and for that Naturphilosophie had to begin with the material provided by the sciences: "not only must philosophy be in accord with experience, the origin and development of scientific philosophy necessarily presupposes and is conditioned by empirical physics."[9] On the other hand, while Hegel did not regard the task of Naturphilosophie as an a priori constitutive determination of empirical science, neither did he regard it as the simple collection and categorization of scientific principles. In Hegel's view the phenomena of nature and the principles of the sciences are tied to one another through immanent connections: that is, they follow from one another by necessity. This is made possible by the fact that, according to Hegel, the categories of logic are not only the structure of human language and consciousness, and thereby sources for the structure of scientific theories, but they are simultaneously the structure of the historical world. Nature is intelligible not because it submits silently to the imposition of an arbitrarily fashioned logical framework, but rather because it is grounded in a concrete logical structure immanently present in the material world. The object of natural science is to reflect that logical structure, while that of the philosophy of nature is to grasp it and raise it to the level of consciousness.

The identification of three traditions of Naturphilosophie opens up fruitful possibilities for exploring science during the Romantic era. Von Englehardt, Snelders, and Löw have already made major contributions to our understanding of the role of each of these traditions in the development of chemistry.[10] An examination of the biological and medical thought of the Romantic era also reveals the presence of these three traditions. In the biological sciences, however, particularly natural history, comparative anatomy, and physiology, two of these traditions were predominant. One tradition was underpinned by the view of nature characterized above as transcendental Naturphilosophie; the other major tradition espoused the metaphysical style and was graced by Hegel's own systematic Naturphilosophie and by the works of Oken, Goethe, and Carus. Speculative Naturphilosophie after 1800 tended, by contrast, to be more confined to medical theory and practice.[11] The aim of the present study is to identify the major practitioners of one of these two schools of biological thought and the characteristic features of its approach to organic nature.

The subject of the present study is to explore the origins and development of the Göttingen School of biology, for it was at Göttingen that transcendental Naturphilosophie had its most significant impact on biology. The distinctive approach to biology practiced by Göttingen biologists derived from ideas fashioned principally by Johann Friedrich Blumenbach during the 1780s and 1790s. Blumenbach's most significant achievement, from our point of view, was to synthesize some of the best elements of Enlightenment thought on biology, particularly aspects of the works of Buffon, Linnaeus, and Haller, in terms of a view of biological organization that he found in the writings of Kant. After discussing the background and evolution of Blumenbach's ideas, I will turn to the further development of this Kantian biological tradition in the writings of several of Blumenbach's students and colleagues, among them Carl Friedrich Kielmeyer, Alexander von Humboldt, and Gottfried Reinhold Treviranus.

The Göttingen School

Although the most important stages in the development of the research tradition that I have identified with transcendental Naturphilosophie were formed in the late 1780s and early 1790s, those developments were prepared in part by institutional arrangements established at Göttingen several years earlier. Not the least significant of these was the organizational planning of the University itself; for unlike other contemporary German (and European) universities that treated the science faculty as a necessary but by no means central part of the university, Göttingen was from the beginning organized around its science faculty, and in the early days around the medical faculty in particular. One simply cannot compare Göttingen with other European universities in the eighteenth century without perceiving the predominant role of the empirical and mathematical sciences at that institution. The reason for this lay partly in the fact that the University did not come into formal existence until 1737, a time at which the sciences were becoming regarded as the necessary basis for the rational and enlightened construction of society.

Perhaps the most important factor in shaping the scientific orientation of Göttingen was the educational vision of its founder and first currator Gerlach Adolph von Münchausen (1698-1770). Von Münchausen was a minister in the Hannoverian regime of George II. Whereas other princes such as Friedrich II had established scientific academies primarily as a personal ornament attesting to their enlightened spirits, von Münchausen had another idea. While he certainly did conceive of the University as an embellishment for Hannover, von Münchausen saw in the formation of Göttingen a means of establishing important political connections and spheres of influence among continental principalities for the fledgling Electorate of Hannover, which had only itself come into existence in 1692. What von Münchausen sought was to create a university that would quickly surpass all others in its reputation for scholarly excellence. The sons of princes and nobility would be attracted to such an institution; it would also serve as a training ground for diplomats and an obvious means for laying the groundwork for future political ties.[12]

There were two important aspects of the plan for catapulting Göttingen into the forefront of European universities. The first entailed a radical new conception of the university and the role of its professoriate. In the prevailing eighteenth-century view, the role of the university professor was simply to provide instruction in the various disciplines. He was expected to be a master of the doctrines in his field, but he was not expected to do research. In a letter to a friend, Johann Mosheim, who later became the first chancellor at Göttingen, provides a description of the typical university professor of his day:

One hour of conversation with a reasonable friend contributes more in my opinion to the advancement of true science than several days lecturing. Moreover there is almost no one here [Helmstadt] with whom I can talk concerning my current research interests. If anyone were to ask the majority of us, what is a professor?, he would be described as a man who is paid for lecturing to young people a couple of hours a day, and who afterwards enjoys himself with friends. Everything centers on sensual pleasure, and that which goes by the name of scholarship is considered secondary and external to the work at hand.[13]

The persons consulted by Münchausen in setting up the general orientation and curriculum of the new university were opposed to this view of things. As a result a university came into being at which the professors had a twofold *official duty;* namely, teaching and research.

In addition to the new research responsibility of the faculty, there was also a shift in the sort of research to be supported by the state. Throughout the Enlightenment, the typical pattern of state-supported research was practical or applied. Science was to be the handmaiden to technology and industry and therein lay its potential contribution to society. Because scientific theory was linked so closely with systematic world-views and theological orientations in the eighteenth century, any attempt to encourage the development of theory ran the risk of supporting a sectarian position.[14] While the founders of the Göttingen educational program certainly were sensitive to this problem, and while they were above all interested in the improvement of society through the practical application of science, they were also strongly committed to the support of pure scientific research.[15] In order to promote research among the faculty members, as well as for the purpose of establishing links to other scientific societies, the *Königliche wissenschaftliche Societät zu Göttingen* was planned as an integral part of the new university from its inception. Members of this prestigious society were required to contribute papers annually. One need only look at the prize questions for the "Historische Classe" to get a sense of the new emphasis on pure research at Göttingen. While other German scientific societies were still promoting the usefulness of historical studies for the questions of legal right and especially for encouraging patriotism, the *Königliche Societät* at Göttingen promoted the study of history for its own sake without reference to its potential use.[16]

Mention was made above of the fact that Münchausen's plan for Göttingen contained two important new aspects. The first was the new role of the professor as researching-teacher. The second concerned the importance of empirical science in the university curriculum. A truly enlightened spirit, Münchausen believed that the persons best qualified to attend to the practical matters of state were those who had developed the skills of logic and a keen sense of observation.[17] The effect this point of view had on shaping the curriculum is unmistakable, for there was a strong empirical component in each of the areas of study. Under the strong influence of Mosheim, courses in history, for example, stressed the use of numismatics and the careful use of archival material.[18] Christian Gottlob Heyne's lectures on Greek literature stressed the importance of archaeological studies for grasping the content and development of Greek mythology.[19]

Perhaps the most important consequence of this emphasis on empiricism for the future development of Göttingen as a center for research was the establishment of the medical curriculum. The decision to exclude all medical theory based on uncertain speculation in favor of doctrines resting on the *terra firma* of careful observation and experimentation led to the formation of a curriculum based on the medical theories of Hermann Boerhaave.[20] In order to strengthen this orientation Werlhof, whose task it was to assemble the medical faculty, sought to attract Boerhaave's most illustrious student, Albrecht Haller, to the University. Haller joined the faculty in 1737.

The specifics of Haller's incredible fifteen-year career at Göttingen need not be recounted here. It suffices to mention that along with his prodigious scientific productivity he was also a crucial figure in establishing links with the rest of the scientific community through the construction of the general organizational plan of the *Königliche Societät der Wissenschaften,* of which he was the first president (1751), and through the establishment of the *Göttingische Anzeigen der Gelehrten Sachen,* to which he contributed at least two hundred review articles annually. It was through the organizational efforts first of Münchausen and later Haller that a new approach to natural philosophy was provided with the institutional structure that it needed in order to develop, a point that will become evident when we consider the effects of Haller's work in Göttingen.

At an institution conceived to foster an interest in pure scientific research, it was not unlikely that in such a supportive environment a new orientation toward the role of hypotheses in theory construction would develop, and indeed such an orientation was first made manifest in the works of Albrecht von Haller, particularly in his thought concerning natural history. Haller set forth his views on hypotheses in science in the introduction to the first volume of the German translation of Buffon's *Histoire Naturelle* in 1750. This essay was printed separately in Haller's *Vermischte Schriften* and was cited in the German literature under the title, "Vom Nutzen der Hypothesen."

Haller did not support Buffon's general plan of natural history. It was far too speculative for his taste. Haller preferred to attempt the provisional construction of systems of nature only after the difficult task of collecting data and constructing experiments was well in progress. Buffon was premature in this regard. Moreover, Haller had grave doubts about the cornerstone of Buffon's entire plan, namely the theory of generation. Nevertheless, even if somewhat tongue-in-cheek, he did see some utility in Buffon's work, particularly since it attempted to bring all of natural history into a single framework. Even if false, such an attempt could lead to a consideration of the real links between scattered areas of research and ultimately to a genuine understanding of the system of nature.

While he regarded the speculative use of hypotheses characteristic of Cartesian science as inimical to scientific progress, Haller did nonetheless see a legitimate role for hypothesis in theory construction. He sided with Linnaeus in considering the distinguishing characteristic of man to be his ambition to master nature, and the tool that enabled him to do so, he argued, was theoretical knowledge.[21] Accordingly, Haller was opposed to the reigning philosophy of science which attempted to ban all use of hypothesis in science.

The new wisdom has it that at some future date all arbitrary opinions, all hypotheses, will be completely banned. . . . [The reason for this restriction] is the assumption that man is prevented from grasping the inner nature of things, that he can at best hope for perceptions of the phenomena, and that the Truth lies beyond a chasm over which he has no bridge.[22]

Unlike most of his contemporaries, Haller did not think that the truth lay for-ever inaccessible to human reason, but he also felt that it was an illusion to think that empirical knowledge could ever be free from all hypothetical com-ponents. He regarded this tendency to eliminate hypothesis from empirical knowledge as an attempt to force a mathematical conception of rigor in a domain where it was not applicable. Furthermore he pointed out that even in recent mathematics, progress rested upon hypothetical foundations regarding the infinite for which no rigorous justification could be given.[23] He observed that some of the most important recent advances in empirical science had been made through the exploration of Newton's hypotheses regarding the aether.[24]

What then was the true use of hypotheses? According to Haller:

They are short of the Truth, but they lead one to it; and I say moreover, man has yet to find a better path, and I can think of no discoverer, who has not himself made use of it. When Kepler wanted to determine the laws of planetary motion, he constructed an hypothesis, an improbable one at that, whose falsity has now been demonstrated; and yet this hypothesis led him to the most wonderful law . . . concerning the periodic times . . . which was firm enough for Newton to build upon.[25]

Thus, "[hypotheses] pose questions whose answers demand experiences which would not have occurred to us otherwise; an effect with untold advan-tages for science."[26]

This discussion served as an introduction to Buffon's *Histoire naturelle.* Although he disagreed with many of the hypotheses upon which Buffon's natural history was constructed, particularly the theory of generation, Haller felt nonetheless that Buffon's theory raised a number of interesting questions worthy of further exploration by German scientists.

There was a second reason for Haller's encouragement of exploration of the questions raised by Buffon that also related to his philosophy of science. Since hypotheses led to the discovery of different aspects of nature, the most profitable path for science was to develop as many complete systems of nature as possible; for out of the unification of these different viewpoints a theory that approximated nature more closely would emerge. This idea that the path toward constructing the true system of nature lay in unifying the greatest multiplicity of different theoretical perspectives became a central feature of later discussions on the philosophy of nature at Göttingen.

Whatever Haller's own reasons for encouraging the study of Buffon's work, the fact of the matter is that a number of aspects of Buffon's *Histoire* were explored by subsequent generations of students at Göttingen. Widespread dissent may have been registered to some of his specific conclusions, but Buffon's *Histoire* was regarded nonetheless as an intellectual *tour de force.*

In his lectures on natural history, for instance, Blumenbach ranked Buffon along with Aristotle, Harvey, Linnaeus, and Haller as one of the major theoretical minds of the subject. One aspect of Buffon's work central to later developments at Göttingen was the insistence on a pluralistic approach in constructing a natural system. Buffon's views on this matter were developed in his criticisms of the systems of Linnaeus and Tournefort, which, he argued, had been based on arbitrarily chosen single characteristics.

Like Haller, Buffon argued that the object of scientific knowledge, though unattainable at present, was to grasp the inner nature of things. Yet this was a knowledge that extended only to the general outlines of nature.

One does not envision that with time we can go so far as to grasp each and every individual thing not only according to its form but also to understand everything that belongs to its birth and genesis, to its internal structure . . . in a word everything that belongs to the natural history of each individual in particular.[27]

Because the natural scientist has no direct intuition of the internal nature of things—or as Kant would phrase it later, because Reason cannot claim to know things in themselves—the only path to knowledge is to construct the *general system* of nature through a comparative analysis.

Insofar as we have no other means to acquire a knowledge of natural things, we must pursue this path as far as it leads us: We must take all things together and gather information concerning their similarities which is useful to us in distinguishing them more clearly and knowing them better.[28]

Latent in Buffon's position was a tension over which the transcendental and romantic Naturphilosophen would later split. Strongly influenced by Leibniz, Buffon believed that once a system had been constructed that harmonized all its elements into a unified view of nature, that system would in fact reflect the real constitutive relations of nature. Certainty regarding the correspondence of this system with nature would be imparted through a kind of intellectual intuition that would emerge through an aesthetic sensitivity to the natural liaison of phenomena thus organized. Kant, by contrast, always insisted that the necessity of seeking systematic unity among theoretical aspects of nature is the paramount demand of reason; but it is a subjective demand that cannot be accorded objective reference. Moreover, only God in Kant's view is capable of the sort of intellectual intuition demanded by Buffon and Leibniz. The human faculty of intuition is sensuous and discursive according to Kant; it can never be intellectual.

Two features of Buffon's comparative method were radical departures from contemporary thought on taxonomy and systematics. The first was his insistence that arbitrarily selected single characteristics such as leaf shape, flowers, or, in the case of the vertebrates, single anatomical characteristics such as claws or teeth, were insufficient means for distinguishing organisms into groups, orders, classes, genera, and species. Such classifications might indeed be useful for certain purposes, but there was no guarantee that they corresponded to natural divisions. The only means of approximating a natural classification was in terms of a system based on multiple characteristics:

One intentionally renounces the greatest advantages which nature offers us unless he makes use of every part of the thing being observed. . . . This is the methodical ordering which must be followed in the classification of natural things: It is important to realize that the similarities and dissimilarities of things are not to be taken from single parts; rather that the descriptive method must be based on the shape, size, and external appearance of various parts, on their number and position, and even on [the chemical constitution of] the material constituting them.[29]

This passage emphasized a major feature of Buffon's approach, later developed more fully by the Göttingen School, namely that the construction of the natural system demanded a classification based not only on multiple morphological criteria but also on systemic physiological critera as well as knowledge of the specific chemical composition of the organism. As we shall see, Blumenbach, Kielmeyer, and Treviranus regarded these different classificatory levels, which they called *Struktur, Textur,* and *Mischung,* as essential to the construction of the natural system. Moreover, a point not evident in the passage quoted above but equally important for his view, Buffon emphasized that a theory of the internal principles of organization must be conjoined with a consideration of the relationship of organisms to their environment as well as to other organisms. This too became the hallmark of the Göttingen program. But it should be noted that by "internal principles" Buffon had something much grander in mind than even the formidable systematic correlation of organic phenomena just mentioned; for in a philosophical sense even these might still be considered "external" characters. What Buffon sought was the essence of animal form itself, the productive source of all these different levels of "external" characters. It was this essentialist dimension of his thought that the Newtonian physiologist Haller had found objectionable, and it was this essentialist dimension of Buffon's philosophy of nature that the Göttingen School rejected. The inheritors of this dimension of Buffon's thought were the metaphysical Naturphilosophen.

This method would produce a series of different classifications just as the so-called "artificial" systems of Linnaeus and Tournefort had done, but this would only be a temporary system subject to constant revision. All systems in Buffon's opinion were necessarily artificial in that they represented at best a series of correlations between *external* characteristics. The object of scientific knowledge, however, was to grasp the essence of the individual. Accordingly, Buffon argued that the various divisions in even the most extensive classification scheme did not refer to things existing in nature: only individuals exist in nature. In order to catch these individuals the natural historian must cast out a net with ever finer systematic gradations. "The more one multiplies the divisions [Einteilungen] among natural beings, the closer he comes to the Truth, because nothing actually exists in nature except individuals. Genera [Arten], classes, and orders exists only in our imaginations."[30]

This distinction between the artificial character of the products of reason on the one hand as opposed to the multiplicity of real individuals existing in nature on the other was a distinction that traced its origin to Buffon's views on epistemology,[31] but it gave rise to a second distinguishing feature of his philosophy of nature. While reason could not immediately attain to an intuition of things in themselves it could nonetheless acquire a dim intimation of what they are like by grasping in a single unified system all the external characteristics of things:

. . . the construction of a general theory requires that everything should be contained within it.[32]

. . . the only true method is nothing other than a complete and correct description of every individual in particular.[33]

As we shall see, two ideas dominated discussions on the construction of a natural system in the works of scientists at Göttingen in the latter half of the eighteenth century. The first was the notion that the only means for attaining a natural system was through a unification of a multitude of "artificially" constructed systems under a single idea; and the second was the related idea that only a complete description of nature [eine vollständige Naturbeschreibung] that united under a single plan the laws of the phenomena from all the various domains could succeed in grasping nature as it is in itself.[34] These were ideas central to Buffon's *Histoire* and while it cannot be said that Buffon was the only source for such ideas,[35] it was through the sponsorship of his work by Albrecht von Haller and others at Göttingen that they occupied a focal position in scientific discussions there.

There was another feature of Buffon's work that became central for later developments at Göttingen, but it concerned an issue over which Haller differed sharply with Buffon. As we have seen in our consideration of the views presented by Buffon in the introduction to his work, a true understanding of individuals contained in a complete descriptive system depended upon grasping the process through which they came into being. Consequently a theory of generation was an essential feature of the natural system. As Buffon saw it, the essential problem was to explain how individuals of the same species reproduce their own kind. That Haller defended a preformationist theory of generation while Buffon proposed an epigeneticist account is well enough known not to require elaboration here. Two features of Buffon's model deserve special consideration, however, for they were important for later developments at Göttingen.

In order to account for the special nature of organic development Buffon postulated the existence of certain elementary organic particles, which he imagined as being probably spherical in shape. The similarities between Buffon's corpuscles and the monads of Leibniz are striking:

It appears highly probable to me that in nature there are innumerable small organic beings that are similar to the large organisms which are manifest [to our sense], and furthermore that these smaller organisms consist of living organic particles which are common to both plant and animal life. These organic particles are elementary and indestructible. A collection of such particles make up the organisms of which we are aware. Consequently generation is merely a change of shape which takes place through the addition of these similar particles just as dissolution of their [spatial] arrangement destroys the whole.[36]

In order to explain how these organic "elements" were shaped into the multiplicity of structures present in nature Buffon took refuge in a construct which he called the *moule intérieur*. The *moule intérieur* was conceived as a kind of structuring force responsible for reproduction, growth, and development, but the mechanism in terms of which it functioned remained unexplained in Buffon's work. In fact Buffon argued that a full understanding of the operation of the *moule intérieur* was not possible. "We will never have a complete correct conception of the characteristics [of the *moule intérieur*], because . . . they are not external characteristics and consequently do not fall within the domain of our senses."[37] For Buffon the *moule intérieur* was the essence of the animal productive of but never grasped in the phenomena to which it gave rise. The only means of incorporating it into an empirical scientific theory was to follow the lead of Newton. Buffon compared the structuring force of the *moule intérieur* to Newton's conception of universal gravitation: in the same way that we must assume that all bodies attract each other in terms of a force whose effects only can be perceived, so must we assume that the *moule intérieur* exists though we may never be able to provide a mechanism that accounts for its effects.

Although Haller left Göttingen in 1753, he continued to exert a strong influence on the development of science there not only through his many personal contacts but also through numerous editions of his works prepared by former colleagues and students, which formed part of the core curriculum at Göttingen.

Four of Haller's former colleagues and immediate successors at Göttingen were instrumental in developing the philosophy of nature found in Haller's own works as well as elaborating some of the themes found in Buffon's

Histoire. One person significant in bringing about further discussion of the philosophical issues raised in the Haller-Buffon controversies was Abraham Gotthelf Kaestner, whose influence in German scientific circles at the time was second only to that of Euler. Kaestner is primarily remembered for his work as a mathematician, particularly in the area of analysis, as well as for having initiated discussion on the provability of Euclid's parallel postulate. Before he came to Göttingen in 1756, however, Kaestner had been the translator of the first three volumes of Buffon's *Histoire,* and in his *Anmerkungen* to Buffon's text Kaestner raised some of the key issues explored later by the Göttingen School.

While openly a strong supporter of the Linnaean method of classification, Kaestner acknowledged that there were unavoidable deficiencies in the so-called artificial system.[38] While he found Buffon's theory of the *moule intérieur* an interesting speculation, he sided with Haller in pointing out that the manner of its functioning remained dark and in need of further clarification.[39]

Kaestner did not teach courses in natural history at Göttingen. His primary teaching activities were in mathematics and physics. Nevertheless it is clear that he continued to exert an influence on questions regarding natural history and on a related topic of increasing interest at Göttingen, *allgemeine Naturlehre.* That Kaestner exerted such an influence is evidenced by his students. Although he wrote his dissertation under Kaestner on matters concerning Euclid's parallel postulate, Georg Simon Klugel later composed a work entitled *Anfangsgründe der Naturlehre,* which falls in the area of philosophy of nature. Stronger evidence of Kaestner's influence is provided by the works of his two most illustrious students, Lichtenberg and Blumenbach. Both of these men acknowledged their debt to Kaestner, and, as we shall see explicitly in the case of Blumenbach, both explored questions raised by Kaestner in his *Anmerkungen* to Buffon's text.

A colleague of Kaestner who taught courses supportive of the general orientation toward the philosophy of nature we have been exploring was David Sigismund Büttner (1724–1768). Büttner's path to Göttingen was indeed an interesting one. Having lost his father at the age of four, Büttner was reared in the home of Georg Ernst Stahl from 1728 to 1735, an experience that may have influenced his scientific orientation considerably. Büttner studied with Haller for one year during 1745 and then moved on to study in Leiden for two years. After teaching chemistry at the Akademie in Berlin for several years, Büttner came to Göttingen in 1760 highly recommended by Leonhard Euler.[40]

Despite the bad press that Stahl has received in historical literature on the development of eighteenth-century medical theory, there is a remarkable affinity between the organic philosophy of nature presented in the lengthy (and repetitive) introduction to his *Theoria medica vera* (Halle, 1708) and the views of the Göttingen School. In that work Stahl set forth a theme we have already encountered in the works of Buffon.

There are two ways in which things can be made the object of research: one can observe them either in their *being* or one can consider them in their process of *becoming;* and with respect to the latter, consideration is to be given to the conditions under which it must occur.[41]

While the first method was purely descriptive in focusing on external characteristics such as size, shape, and quantitative aspects, the second approach concerned itself with the internal nature of the things themselves. Rather than constructing a general conceptual framework of nature, Stahl claimed that the true task of science is to grasp the particular properties of individuals.[42] Such an understanding of things would only be forthcoming, however, with a complete cosmology that demonstrated the nexus of relations linking all individuals.

We do not want to hunt hastily for the universal and final goal according to which all things are constructed and ordered, . . . much less do we want to deceive ourselves through arbitrary fantasies. In order to protect ourselves from all self-deception we ought to consider the teleological relationships in which individual things stand to one another, and accordingly we should concentrate on an historical consideration of these relations through the proper use of the understanding.[43]

Stahl, who was strongly opposed to the tendencies of the mechanical philosophy, proposed a return to an Aristotelian organic philosophy of nature that took as its fundamental datum the unmistakable fact that nature strikes us as *organized,* as constituting a structured whole or Cosmos. According to Stahl's holistic philosophy, the laws of mechanics should be considered a subset of the laws of organic nature.[44]

That Stahl's views were taken seriously at Göttingen in spite of Haller's opposition is evidenced in the works of Blumenbach, who regarded Stahl as "one of the deepest thinking physiologists."[45] The extent to which Stahl's organic philosophy of nature entered Göttingen through Büttner's influence is difficult to determine, however, for Büttner died at an early age having left as his only publication a work on the polyp, which appeared in the *Proceedings of the Royal Society* in 1752. There is some evidence of his teaching activity at Göttingen between 1760 and 1761, however, which sheds some light on his general intellectual orientation. In his history of the university, Johann Stephen Pütter described the teaching activities of his young colleague. Büttner normally taught courses on chemistry and Materia medica, but in the summers he taught a special course devoted to the medicinal plants which he ordered according to their natural affinities, . . . since then the powers of such plants and their effects on the human body could be explained. He makes a special effort to predetermine the inner essences [innerliche unsichtbar Kraftwesen] of these plants merely from botanical similarities and relationships [verwandtschaften] of the plants among one another as well as from their external characteristics which immediately strike the senses.[46]

While such a course need not have been the special preserve of one committed to an organicist philosophy of nature it did contain a number of features in common with the philosophies of nature of Stahl and Buffon. There was a concern for grasping the inner essence of the things themselves coupled with a methodological prescription that this could only be accomplished through a complete description of all the external characteristics of the plant under consideration and its relationship to the rest of nature. As a first step in grasping the *innerliche Kraftwesen* a classification of external characters was to be constructed.

A similar orientation was manifest in the work of Christian Wilhelm Büttner, a colleague not related to David Sigismund Büttner. C. W. Büttner was active at Göttingen for twenty-five years from 1758 to 1783. He taught a variety of courses including natural history and chemistry. His natural history collection, which had been begun by his grandfather, Ulrich Büttner, was famous throughout Germany. Part of it served as the basis for the Museum of Natural History at Göttingen, although Büttner later (1783) gave the major portion of the collection to the University of Jena in exchange for a sinecure.

A number of his students, including Blumenbach,[47] acknowledged C. W. Büttner as having exercised a formative influence on their thought. Two aspects of his orientation toward natural philosophy seem to have been especially important. Like others in the tradition emerging at Göttingen, Büttner considered the true path to knowledge to consist in grasping the "inner nature" of things. Pütter, for example, reports that "in his chemical lectures Chr. Wil. Büttner seeks to show in particular that chemistry can serve as a key to the knowledge of the innermost nature of things themselves."[48]

Along with the interest in grasping the nature of individuals, we have seen a developing interest in extracting a general system from a catalogue of phenomenal relations and characteristics. Büttner's work manifests a similar tendency. A special feature of his course on natural history was the inclusion of man as an object of study. Büttner was particularly interested in the development of civilizations and he focused on the study of language as a primary indicator of development. Though obscurely formulated in his writings, Büttner advocated a method that later became a central feature of the work of another Göttingen student with strong leanings toward metaphysical Naturphilosophie, Wilhelm von Humboldt. Büttner proposed that the languages of various cultures all be considered as manifestations of a single *Urform* that had been expressed differently under different environmental conditions.[49] Of course, such ideas had a long tradition stretching back to the Hippocratic work, *Airs, Waters, and Places,* but the need for constructing an *Urbild* or Ideal Type was typical of works at Göttingen during this period that explored the issues surrounding Buffon's *Histoire naturelle.*[50]

The notion of an *Urbild,* or *Urtyp,* played a central role in the work of another Göttingen professor, Christian Gottlob Heyne, who was the teacher of Blumenbach as well as a number of prominent Naturphilosophen. A theory of art criticism bearing the marks of views expounded by Winkelmann, but more importantly by Diderot in his *Essais sur le peinture,* formed the core of Heyne's course of lectures on archaeology at Göttingen.[51] In Heyne's view the central task of archaeology is to analyze the artistic creations of the ancient world. In order to do this he recommended that the student must ultimately immerse himself in what he called the *total habitus* of the civilization under consideration. To grasp the significance of some artistic production within a culture, a statue of Apollo for example, the archaeologist must be exposed to every empirical aspect of the culture, including the climatic variations of the region, in order to penetrate the true spirit of that civilization. Having been so prepared, he must finally construct an ideal type.

The artist can take numerous paths in order to construct the ideal. He can do it through the imagination or through the interrelation of individual parts into a whole. The parts of this ideal form will never be found in nature itself, rather they will be scattered here and there.[52]

As in the works of other Göttingen professors we have examined there was an emphasis on grasping the individuals encountered in experience under the unity of a single plan. This plan emerged from a thorough immersion in all the relationships between individuals in the total habitus. In Heyne's theory, reason grasped the *Urbild,* or ideal type, in the same way that Buffon had argued that it grasps the *moule intérieur,* and in the same way that Diderot argued that it grasps the *"beau ideal"* in art: namely through an intuitive act of the imagination that gives rise to a perception of a "secret liaison," a necessary connection in all the differences between individuals.[53]

From the preceding discussion we can see that at Göttingen in the 1750s and 1760s the elements of a distinctive philosophy of nature were beginning to form. As yet no clear statement of that philosophy had been put forward, but in the series of questions surrounding Buffon's work and in the general intellectual orientation of key figures at Göttingen during this period the outlines of what would emerge as a fully articulated system in the late 1780s and early 1790s can be seen.

The issues from which the new philosophy of nature would develop centered on natural history, and in particular on the construction of a natural system. At the heart of discussions of the natural system was the belief that only a theory that comprehended all existing individuals under the unity of a single plan would suffice. Such a theory would depend necessarily upon a grasp of the "internal nature" of things, on the *Urform* that underpinned their phenomenal existence, and on the environmental conditions responsible for its particular manifestation. As the problem came to be formulated in Göttingen scientific circles such a theory would show how "Alles in allem verwebt ist." In the wake of sceptical critics such as Descartes, Locke, and Newton, and in the mounting onslaught of Berkeley and Hume, such a philosophy of nature stood in desperate need of an epistemological critique that solved the riddle of how human reason, confined in its operation to the combination of "external" characteristics derived through sense perceptions, could ever penetrate the *innerliche Kraftwesen* of things. One potential solution visible in the early writings of Göttingen natural philosophers such as Kaestner was to argue that reason simply does not have access to such internal unities. The best it can do is seek out laws governing different aspects of organic phenomena and then seek to relate them into a synthetic whole, perhaps through establishing systematic interconnections in terms of a single natural force or small set of mutually dependent forces. This was the approach ultimately developed by the Göttingen School as a result of the influence of Blumenbach and his acceptance of Kant's views. Another potential solution to this problem, also visible in the writings of the Göttingen professoriate, was to seek a theory of mental activity, at once sensuous but also intellectual, that permitted access to the source of the "secret liaisons" of phenomena. This approach was favored by those who, like Diderot, saw in aesthetic theory a solution to some of the most difficult epistemological questions confronting the investigation of nature. This latter approach was explored by the Romantic Naturphilosophen.

Laying the Theoretical Foundations of the Göttingen School: The Union of the Teleological and Mechanical Systems of Nature in the Works of Kant and Blumenbach

Although there was a common set of themes in what I have attempted to describe as a shared general orientation to natural philosophy developing in the courses and writings of several Göttingen professors in the period after Haller's departure, there was, as yet, no attempt made to synthesize the various elements of this "orientation" into a coherent view of nature. Such a synthesis was first approached in the work of Johann Friedrich Blumenbach (1752-1840). Blumenbach began his studies at Jena, but was encouraged by his father, who was a close friend of Christian Gottlob Heyne, to move on to Göttingen. Blumenbach matriculated at Göttingen in 1772.[54] He studied under both Heyne and C. W. Büttner, both of whom he acknowledged fondly in his dissertation *De generis varietate humani nativa* of 1775. Blumenbach was appointed to the faculty in 1776 and became professor of medicine in 1778. Through his incredible scientific productivity, nearly equal to that of Haller and Kaestner, and through a long and active teaching career extending over nearly half a century, Blumenbach became one of the most influential German theoreticians on questions of natural history in the late eighteenth and early nineteenth centuries.

In his dissertation *De generis varietate humani nativa* Blumenbach was already at work synthesizing themes from the discussions of his predecessors at Göttingen that would form the basis of the philosophy of nature that emerged in the late 1780s. While most writers on the subject regarded the different varieties of man as clearly distinct species, the aim of Blumenbach's dissertation was to show that there exists only one human species and that the various races are "degenerations" of a primary species or Stammgattung.[55]

At the heart of the problem of determining the characteristics of the Stammgattung was the thorny issue of identifying true species and distinguishing them from races. As Blumenbach noted this was no simple matter:

What is *species*? We say that animals belong to one and the same species if they agree so well in form and constitution that those things in which they differ may have arisen from degeneration. . . . Now we come to the real difficulty, which is to set forth the characters by which *in the natural world* we may distinguish mere varieties from genuine species.[56]

Two paths lay open to him: that of Linnaeus and that of Buffon. As the definition above indicates he preferred the Linnaean method of referring to morphological characters in determining species. Blumenbach argued that a determination based on the capacity to interbreed and produce fertile offspring, while undeniably a true criterion for distinguishing species, was insufficient for the purposes of empirical investigation. Although wild species do indeed reproduce fertile offspring only with members of the same species, Blumenbach reasoned that the application of this as a criterion for distinguishing species was possible for only a few cases. One would like to know, for example, whether the Indian and African elephants are races of the same species. Answers to such questions could in practice only be given by an anatomical investigation.[57]

Blumenbach's rejection of Buffon's breeding definition for species did not imply an unconditioned acceptance of Linnaean principles of taxonomy, however. Rather, Blumenbach followed the lead of others at Göttingen who had agreed with Buffon's arguments in the General Introduction to the

Histoire naturelle which stressed the importance of multicharacters for determining the natural system.[58]

Blumenbach proposed to construct a system based on the *total habitus* of the organism under consideration. This would include not only a determination based on multiple anatomical characters but would also include the relationship of particular forms to the environment. Once the stem for each species had been identified Blumenbach intended to provide a theory accounting for the causes of their variations. And in setting out this theory he vowed to follow Newton's rules for philosophizing about nature:

As we enter upon this path we ought always to have before our eyes the two golden rules which the great Newton has laid down for philosophizing. First that the same causes should always be assigned to account for natural effects of the same kind. We must therefore assign the same causes for the bodily diversity of the races of mankind to which we assign a similar diversity of body in the other domestic animals which are widely scattered over the world. Secondly, that we ought not to admit more causes of natural things than what are sufficient to explain the phenomena. If therefore it shall appear that the causes of degeneration are sufficient to explain the phenomena of the corporeal diversity of mankind, we ought not to admit anything else deduced from the idea of the plurality of the human species.[59]

Having adopted this methodology Blumenbach addressed the question, "What is it that produces now a better now a worse progeny, at all events different from its original progenitor?" A first attempt proposed only to be immediately and unconditionally dismissed was that degeneration could occur through fertile hybridization leading to an ultimate transformation of species.

The latter case [fertile hybridization] although rare (and that by the providence of the Supreme Being lest new species should be multiplied indefinitely) I would admit of in beings closely allied. . . . But from all this we must carefully separate the plainly fruitless unions of animals of different species. . . . There are good reasons for refusing to believe that from any incongruous attempt . . . offspring can be born or even conceived. First to consider is the unequal proportions of the genital parts in many which are providentially and carefully adopted for copulation in either sex of the same species, but in distinct genera render the entire thing impossible. [Here B. cites Haller, Physiologie, Bd. viii, p. 9] . . . Besides . . . in each species of animal there are certain periods of gestation and pregnancy of the mother, the formation and progressive development of the fetus.[60]

On the basis of the position elaborated in this passage, which he supported by extensive anatomical evidence, Blumenbach went on to conclude that the Ethiopian race could be neither a degenerate form of ape nor a hybrid simian-homosapiens form.

Having eliminated transmutation as a possible mode of degeneration Blumenbach went on to describe what he considered the true cause of the variation of species. The principal source he reckoned to be climate:

There is no diversity of the *habitus,* which may not be produced by varieties of climate. . . . Thus if European horses are transported toward the east, or to Siberia or China, in the process of time they . . . dwindle, and become smaller in body so that you would scarcely recognize them as being of the same species.[61]

European cattle, on the contrary, produce taller offspring under the same conditions. In addition to climate Blumenbach singled out mode of life and differences in the constituents of nutrition as causal agents producing variations in the original *Stammgattung.* Using this theoretical framework he went on to conclude that the Caucasian race was the original stock of the human species and that all other races were degenerations of this original.

While it would be misleading to see all of Blumenbach's later scientific concerns as having been generated by questions raised in his dissertation, still, in my opinion, it is not far from the mark to see the core elements of his later philosophy of nature in this youthful work. Blumenbach prepared two other editions of the treatise (1781, 1795), each one of which was expanded to include the latest results of his own research along with a synthesis of materials drawn from an extensive and careful search of the scientific literature. As Blumenbach noted in the introduction to each new edition of the treatise, he had raised questions in his dissertation for which he could not then provide adequate answers, and the arguments presented therein lacked a solid foundation of empirical support. Many directions of his later scientific career were aimed at providing that needed support.

One element present in the *De generis varietate humani nativa* that became an abiding feature of Blumenbach's philosophy of nature was the notion that classification must proceed by the examination of multiple characters. All twelve editions of his extremely influential *Handbuch der Naturgeschichte* (1st. ed., 1779) as well as a more speculative work, *Beyträge zur Naturgeschichte* (1st. ed. 1790, expanded in 1806-1811), contain clear statements of the importance of multiple characters for constructing a natural system. It is important to note, however, that Blumenbach considered the use of multiple characters as a methodological strategy and that the system constructed in terms of this strategem was still an *artificial* system. In this he took issue with Buffon who, according to Blumenbach, had violated the *methodological* nature of this strategem by inferring from it the *existence* of an all-encompassing network of organisms, a great chain of being:

All the beloved pictures of chains, ladders, nets etc. in nature certainly do have an unmistakable usefulness for methodology in the study of natural history, since they give the basis for the so-called *natural system* according to which all creatures are ordered according to the most numerous and most evident similarities, that is, according to the *total habitus* and the affinities based upon it. But it is a great weakness to see in such pictures the Plan for creation, that "nature makes no leaps" as the natural theologians are wont to say, simply because according to their *form* organisms follow upon one another in a finely graded series of steps [*Stufenfolge*].[62]

The chain of being, he went on to argue, is simply an artifact of the method of classification. Closer inspection reveals unmistakable gaps for which there are no imaginable bridges, as for example between the organic and mineral realms.

A further aspect of Blumenbach's theory of the natural system present in his dissertation but which became the subject for extensive subsequent development was the notion of the ideal type. In the dissertation the various races of man were all described as phenomenal manifestations of an original *Stammgattung* which had been subjected to different environmental pressures. While Blumenbach identified this original *Stamm* with one of the presently existing races, namely the caucasian race, implicit in his treatment of the problem was the idea that the *Stamm* need not exist, that it may have become extinct in one of the revolutions of the globe and that accordingly a number of different organisms might mistakenly be identified as separate species when in fact they were simply races of the same (extinct) *Stammgattung*. Here the systematist was confronted with one of those gaps in the "net" of nature, and it was his special task to provide a hypothetical reconstruction of the *Stamm.*

In order to construct the *Stamm* Blumenbach followed the lead given by Haller in comparative anatomy and physiology. By the second edition of his *Handbuch der vergleichenden Anatomie,* Blumenbach was "daily becoming more convinced that comparative anatomy is the living soul of natural history."[63] It was through comparative anatomy that the ideal type, or *Stamm,* was to be identified. Blumenbach's theory of the ideal type was identical to that found in the works of his teacher (and later father-in-law) Heyne. In his *Geschichte und Beschreibung der Knochen des menschlichen Körpers,* Blumenbach cited as two additional sources for his notion of the ideal type works by Caspar Friedrich Wolff and Denis Diderot. In his paper, *De inconstantia fabrica corporus humani* (1778), Wolff argued that the main problem confronting the naturalist is the variability encountered in comparative studies of individual anatomical parts of the human body:

I do not deny that there are conformities in the human structure and that these are perpetual, . . . but the inconstant characteristics are so mixed and confused with the constant ones that I do not believe there is anyone who can distinguish and define them.[64]

In order to solve this problem Wolff concluded: "Therefore in both the external shape of the human body as well as its inner structure it is necessary to construct a beautiful form."[65] This "beautiful form," though not to be encountered in any individual, was to be abstracted from extensive comparative anatomical studies.

In his "Essais sur la peinture" Diderot also stressed the importance of constructing through an act of the imagination a "beau ideal." While Blumenbach was convinced of the importance of constructing an ideal type, in none of his works did he support Diderot's view that it was to be arrived at through an aesthetic judgment. Instead Blumenbach was more intrigued with another aspect of Diderot's essay. While it was universally admitted that the variability in nature could only be made intelligible in terms of ideal types, Diderot argued that even in its variations nature follows definite rules:

Nature does not do anything incorrectly. Every form whether it be beautiful or ugly has its cause, and of all the things that exist there is not one that should be otherwise than it is. Look at that woman who lost her eye during her youth. The successive increase of the orbit no longer distends the eyelids; they have retreated into the cavity hollowed out by the lack of the organ; they have become smaller . . . the alternation has affected every part of the face rendering them longer or shorter as a result of the accident.
. . . We say that the man passing in the street is misshapen. Yes, according to our impoverished rules, but according to nature it is otherwise.[66]

What interested Blumenbach was the suggestion that one could work back to the ideal type by constructing the laws of its variation. Characteristic of Blumenbach's approach is that ideal types are theoretical constructs. They are a hypothetical synthesis of laws related to various aspects of animal organization. In applying these ideas to his own work, Blumenbach noted that there is always a homogeneous relationship between the parts of the skeleton. If a particular part of a skeleton was found to be well-formed, he claimed to have found that the other related parts would also be well-formed. Moreover, if a particularly important structure was found to be misshapen in some aspect, Blumenbach observed that the other related parts would vary from the *Muster* or *Vorbild* proportionately in the same degree. How these degrees of variation were to be measured he did not say. That he was intent upon finding a quantitative determination, however, is indicated by the fact that he called this relation the *Law of Homogeneity* and claimed that it is the anatomist's most important tool.[67] Moreover, he did not intend to leave this "law" as a mere conjecture. He undertook a study of monstrous births in order to discover the rules underlying variation from the ideal type. The study of misbirths always occupied a central role in his later investigations; he kept a special notebook[68] on the subject up until the last years of his active research (1833); and it always played a central role in the theoretical portions of his works on natural history.

Another feature of the passage quoted from Diderot above came to occupy an extremely important function not only in Blumenbach's approach to biology but it came to be characteristic of the biology of the entire Göttingen School and even of the Romantic era in general. It concerned the degree to which biological science must ultimately be grounded in a *teleological* form of explanation. What so impressed Diderot and Blumenbach is that even when it begins with partially defective materials, nature always strives to bring about the most perfect form possible. The norm followed in this instance is not some abstract concept of beauty; rather it is the production of a functional whole organism given the limitations of the materials from which it is organized and the demands placed on it by the external environment. Birth defects were of enormous importance not only because they revealed the sort of variations an organic system was capable of assuming. Blumenbach and many of his colleagues and students, such as Sömmering, Reil, J. F. Meckel, and Kielmeyer, saw in monstrous births certain definite lawlike patterns. These could not be the result of blind chance. Instead they revealed that at the basis of organic phenomena are forces that do not at all follow the same patterns of cause and effect found in the inorganic realm. They are more self-regulating, striving to achieve some definite end, even when faced with obstacles. By studying the laws of these variations, biological science would attain a fuller understanding of these forces of organization and regulation.

In Blumenbach's work what might best be characterized as a Newtonian research program for natural history was emerging from a consideration of problems surrounding classification given the immense variability in nature. Like Buffon, he was willing to admit that only individuals exist in nature; but these individuals were formed according to a plan, a definite *Urbild* lying hidden in their inner being. The problem was to get at that *Urbild* and to relate it in a system with other *Urbilder.* The goal of reconstructing the *total habitus,* although methodologically useful, did not ultimately provide a satisfactory solution to the problem, because it furnished at best a multitude of *external* characters that would always be ordered in different ways according to the intentions of the systematist. The natural system could only be constructed by penetrating the inner core of the organism, by observing its construction from the inside out, as it were. As Blumenbach told his students in the section of his lectures on comparative anatomy entitled "Allgemeine Übersicht der Geschichte der Anatomie," the true nature of an organism is given by its internal form [nach der innern Form.] [69]

The problem, however, was to get at the inner form when according to the prevailing epistemology, reason could not get at things in themselves. The answers provided by Buffon, Diderot, and Caspar Friedrich Wolff were not satisfactory. Wolff proposed that reason abstract an ideal type from the multiplicity of individuals without providing a theory of mental activity that could guarantee the results. Buffon and Diderot on the other hand seemed to attribute to Reason a mysterious power of grasping the "secret liaisons" between things. But as Blumenbach's teacher Kaestner had pointed out in his note to the section of the *Histoire* on the *moule intérieur,*

Does not Herr Buffon simply conclude that because there are internal characteristics of bodies so must there be an internal form? And is this conclusion so convincing since he has not given a general explanation of what he calls the internal form, rather has he not left us to construct the inner form from the comparison of external characters, which he readily admits apply only to the exterior of things? Does he say, therefore, anything more than that there exist things for the inner nature of a body which are for it what its external characteristics are for us?[70]

Blumenbach believed that it was necessary to construct the *moule intérieur* in a manner that satisfied the requirements of a scientific explanation and at the same time avoided the use of a kind of mystic vision in arriving at its fundamental concepts. This he proposed to do in the manner of Newton: Reason could not penetrate the inner nature of things to provide a causal mechanism for their operations, but it could prove the existence of such beings and the modes in which they appear. The problem must be approached as one approaches problems in perturbation theory: from the observed variations in motion to determine the mass of the unobserved planet and construct its orbit. In terms of natural history this meant that each type of organism had its organizing form. In order to grasp the nature of this form one must observe the patterns in variations produced by environmental factors, in monstrous births and in hybridization. Though never accessible in itself, the inner form was to be treated as a "force" in the Newtonian sense, and the properties of that force would be revealed through the development of structure of which it was the cause.

There was a precedent in this Newtonian approach to the problem in the work of Albrect von Haller, with whom Blumenbach was in close correspondence.[71] In his theory of muscle contraction Haller had argued that the power of irritability, the ability of some parts to contract spontaneously when subjected to an external stimulus, must be considered the effect of an inborn force [*angeborene Kraft*] for which no further explanation could be given.[72] Blumenbach sought to follow the path marked out by Haller in the use of Newtonian-type forces as causal agents in physiology by constructing a force model for the *moule intérieur* that would account for the variety within species in a lawlike fashion. He called this force the *Bildungstrieb* and it was the causal agent responsible for all generation, reproduction, and nutrition.

The theory of the *Bildungstrieb* was first presented in a short paper which appeared in the *Göttingenschen Magasin der Wissenschaften* in 1780. A fuller version appeared in 1781 titled *Über den Bildungstrieb und das Zeugungsgeschäfte.* As the following passage from the *Handbuch der Naturgeschichte* indicates, Blumenbach conceived the *Bildungstrieb* as resolving questions for natural history that Buffon had raised in regard to the *moule intérieur:*

The specific form and habitus of each individual species of organized body is maintained through the determinate, purposeful [*zweckmassige*] effect of the *Bildungstrieb* in the organic materials which are specifically suited to receive it.[73]

Blumenbach was always careful to distinguish the *Bildungstrieb* from anything resembling a soul superimposed upon matter, a *vis plastica* or the *vis essentialis* of Caspar Friedrich Wolff.

The term *Bildungstrieb* just like all other *Lebenskräfte* such as sensibility and irritability explains nothing itself, rather it is intended to designate a particular force whose constant effect is to be recognized from the phenomena of experience, but whose cause, just like the causes of all other universally recognized natural forces, remains for us an occult quality. That does not hinder us in any way whatsoever, however, from attempting to investigate the effects of this force through empirical observations and to bring them under general laws.[74]

Fashioned in the language of the General Scholium to the *Principia,* this passage revealed Blumenbach's goal of doing for organic bodies what Newton had accomplished for inert matter. For each class of organized beings there was a specific *Bildungstrieb* that gave rise to its determinate structure. And just as Newton had succeeded in finding the universal organizing force of inert matter by constructing a model that successfully unified Kepler's laws, Galileo's law, and a host of other "observed" regularities under a single plan, so it was the task of the naturalist to reconstruct the *Bildungstrieb* for each class of organism by unifying the regularities found in reproduction, generation, and nutrition under a general law.

The treatise *Über den Bildungstrieb* proposed a solution to a problem that had been current in Göttingen scientific circles for a generation. Moreover it built firmly on ideas central to the discussions of the previous generation, particularly those of Albrecht von Haller. Indirect evidence from the treatise itself indicates that the theory presented there grew out of discussions with some of the principal figures mentioned earlier. Blumenbach explicitly acknowledged Haller's direct influence on the theory.[75] While Haller had defended the preformationist view in the 1750s in his dispute with Buffon and later with Caspar Friedrich Wolff, Blumenbach believed that he had begun to reverse his position in the 1770s in the latest edition of his *Grundriss der Physiologie,* and this lent great authority to Blumenbach's project. Others mentioned as having been involved in the discussion leading to the production of the theory were Büttner and Lichtenberg.

As we shall see, the treatise *Über den Bildungstrieb* not only brought together a number of key ideas from the previous generation, it also served as a central text around which the ideas on natural history of the generation of students trained by Blumenbach were organized.

The basic model for the *Bildungstrieb* grew out of Blumenbach's experiments on the polyp. What was particularly striking about that organism was not only that it could regenerate amputated parts without noticeable modification of structure but that the regenerated parts were always *smaller* than their originals.[76] Upon closer inspection this seemed to be characteristic of the reproduction of injured organic parts generally. In cases of serious flesh wounds, for example, the repaired region was never completely renewed but always contained somewhat of a depression. Such observations led to two conclusions:

First that in all living organisms, a special inborn *Trieb* exists which is active throughout the entire life span of the organism, by means of which they receive a determinate shape originally, then maintain it, and when it is destroyed repair it where possible.

Secondly that all organized bodies have a *Trieb* which is to be distinguished from the general properties of the body as a whole as well as from the particular forces characteristic of that body. This *Trieb* appears to be the primary cause of all generation, reproduction, and nutrition. And in order to distinguish it from the other forces of nature, I call it the *Bildungstrieb.*[77]

As the name indicates, the *Bildungstrieb* or "formative drive" was considered the force responsible for producing organic structure, and it was conceived to be manifest in three functions specifically associated with the production, maintainance, and repair of structure. As Blumenbach phrased it: "in other words nutrition is a general but continual, reproduction on the other hand a repeated but partial, generation."[78] This way of phrasing the problem makes it clear that the model for the *Bildungstrieb* was to be constructed from careful observations on the generation of animals.

The *Bildungstrieb* was not a blind mechanical force of expansion that produced structure by being opposed in some way; it was not a chemical force of "fermentation," nor was it a soul superimposed upon matter.[79] Rather the *Bildungstrieb* was conceived as a teleological agent with its antecedents ultimately in the inorganic realm, but which was an emergent vital force. This aspect of Blumenbach's work was its distinguishing feature, and it was in terms of this extremely important idea that German philosophers of nature saw for the first time a means of uniting the teleological and mechanical systems of nature.

That the *Bildungstrieb* was conceived as intimately linked to a material basis can be seen from the manner in which Blumenbach claimed to have been led to the idea; namely, that while the polyp always regenerates a lost part, the regenerated part is always smaller. Having lost a substantial portion of its primary generative substance, the force of the *Bildungstrieb* had been weakened. Though its force could be diminished, if it had sufficient strength it would always bring forth the whole structure associated with it.

No small evidence in support of the worth of this theory of the *Bildungstreib* consists in the fact that the shape and structure of organic bodies is much more determinate than either their size, length or other such qualities. . . . not only in the case of water plants but also in the case of animals and even man the size of many parts, even the most important tissues of the stomach and the brain, and the length of the intestines can vary enormously, while the variation in their structure and organization is seldomly ever encountered.[80]

The direction in which this idea would lead can be seen from a section of the copy of the *Handbuch der Naturgeschichte* that Blumenbach used for his lectures. There we find the following:

In comparison with presently existing organisms we find many, even among the pre-Adamistic conchylien, which are identical to present forms. We find others, however, which are *similar* to present forms but which differ substantially from them in size, distinguished partly through small yet constant divergencies [*Abweichungen*] in the formation of individual parts; but also distinguished by the fact that they agree more or less with *Urbilder* that are native only in tropical regions far from the location of the fossils.[81]

When the ideas in this passage are juxtaposed with an aspect of Blumenbach's thought discussed earlier, we see the great potential of the theory of the *Bildungstrieb* for natural history. In the *De varietate generis humani nativa* Blumenbach had said that he accepted as a rare but possible occurrence fertile hybrid crosses between individuals of closely related genera [*Geschlechter*]. Central to the theory of the *Bildungstrieb* was the notion that as the vital force was diminished, smaller but *similar* structures were produced. Thus the largest forms of a particular type might flourish under a specific habitus, the tropical zone for example, while smaller organisms having the same essential structure would be the result of the activity of the *Bildungstrieb* in an environment that diminished its total effective force. Other modifications were also conceivable. By substantially altering the material constituents upon which the *Bildungstrieb* was dependent for the manifestation of its active force, effects analogous to the production of monstrous births might be introduced; there would be a structural modification among classes of forms related to an ancestral ideal functional arrangement of the forces constituting the animal. In short, species falling within a genus could be viewed as "degenerations" of an original stem produced through the agency of the environment. This alteration in the force of the *Bildungstrieb* would also account for the formation of races from the degenerations of a *Stammgattung*.

That such implications were present in the theory did not escape Blumenbach's notice. He drew explicit attention to them in Part I of the *Beiträge zur Naturgeschichte:*

Almost every paving stone in Göttingen bears witness to the fact that species— even entire genuses—of animals must have perished. . . . The structures of an enormous number of fossils in our vicinity are so divergent from all present forms that hopefully no one will seriously attempt any longer to search for them among present forms of life.[82]

Once Blumenbach had committed himself to the view that whole systems of organic forms had been destroyed, the problem immediately thrust forward was the relationship between the forms of the *Vorwelt* and presently existing forms. Had there been several creations or was there a genetic relationship between the forms of the *Vorwelt* and presently existing forms? Had there been several creations or was there a genetic relationship between the old and new orders? Blumenbach took the side of continuity in the operations of nature: "After the organic creation of the pre-Adamistic period . . . had been destroyed by a total catastrophe . . . the Creator allowed the same natural forces to operate in bringing forth the new organic realms."[83] At first glance this move seemed to introduce more problems than it actually resolved; for if entire genuses had been destroyed in the revolutions of the globe, there would seem to be no basis for a continuity of *forms* between the old world and the new.

It was at this point that the notion of the ideal type and its associated *Bildungstrieb* entered the picture.

In order that the formative nature reproduces organisms of a similar type to those of the *Vorwelt,* but organisms which are nonetheless more suited to the other forms in the new order of things, it is necessary that forms be permuted [*hat vertauschen müssen*] since they [are produced] by modified laws of the *Bildungstrieb.*[84]

Two features of this remarkable hypothesis are important to note. The first is that Blumenbach considered the history of nature to consist in a succession of *systems* of interrelated forms. Moreover, the interrelation between forms is a *dynamic* one, the specific characteristics of an organism depending on its relation to other organisms and to the environment. The second point to consider in this passage is that beneath the series of forms in the successive systems of nature is a substrate of permanent Types: forms that are capable of different phenomenal manifestations depending on the conditions in which they are placed. The *Bildungstrieb* associated with each of these Types and the direction it receives from the environment is responsible for the specific differences in the series of forms of the same type in the fossil record.

One further aspect of the theory is especially important to bear in mind. The Types mentioned by Blumenbach are not to be identified with "species." Species, in Blumenbach's view, can be created and destroyed, while the Types of which they are the phenomenal manifestations are permanent. The Types remain,

only the *Bildungstrieb* is forced to take on a more or less altered direction even in the production of new species [*Gattungen*] as a result of the modification of matter arising from such a total revolution.[85]

In commenting on this aspect of his theory Blumenbach underscored that the sort of alteration of form he had in mind was not a degeneration [*Ausartung*], but rather a "modification" [*Umschaffung*] resulting from the changed direction of the *Bildungstrieb*.[86]

The mechanism for such a development was implicit in the materialistic basis of the *Bildungstrieb*. This point came forth most explicitly in Blumenbach's explanation of generation. He ushered numerous reasons for rejecting the preformationist doctrine, but the most significant fact against it in his opinion was that a considerable amount of time was required between fertilization and the appearance of any structured development. During this time even the most careful microscopic investigation could reveal no structure. Once development did begin, however, it proceeded with incredible rapidity. In fact Blumenbach attempted to state a law expressing the rate of development; to wit, that increasing development is inversely related to time.[87] The explanation offered for this phenomenon was that it took time for the seminal material contributed by both parents to "organize" and develop the inherent *Bildungstrieb* which would give rise to structured development.

The specific time associated with the onset of development in each species . . . is explained as soon as it is assumed that the fluids contributed by each parent for generation, the raw materials of the future new organism, require a specific preparation period for their mixture and inner connection and other changes; in a word they require time for ripening before the *Bildungstreib* in them can be excited and the formation of the unstructured material can begin.[88]

The force of the *Bildungstrieb* lay somehow inextricably bound with the constituents of the generating liquid. According to the Newtonian force imagery underlying the model, the mixture of two parent stocks of very different species would cancel each other out:

the mixture of generational fluids [*Zeugungssäften*] of two completely different kinds normally smothers and destroys any disposition for the *Bildungstrieb* which would otherwise be excited. Consequently the possibility of hybridization [*Bastardzeugung*] is limited to very few cases due to the confusion that would necessarily accompany it.[89]

An example of successful transformation by hybridization was provided by Kohlreuter's experiments on tobacco plants. Here fertile hybrid offspring were produced because the *Zeugungssäfte* of *Nicotiana rusticana* and *Nicotiana paniculata* were so closely related. But after several generations the greater "force" of the *Bildungstrieb* associated with *paniculata* manifested itself erasing any trace of a hybrid origin.[90] The same model explained monstrous births. In this case an extremely strong external force or the mixture of two incompatible *Zeugungssäfte* diverted the *Bildungstrieb* from its normal course. Milder but continuous external pressures, however, such as gradual changes in climate either through physical alteration of the environment or through migration as well as changes in the constituents of nutrition, could divert the *Bildungstrieb* from its normal path resulting in the production of races and varieties. All classes of degeneration were thus accounted for in terms of the force produced by the generational fluids contributed by the parents.

Blumenbach had the idea of constructing an entire physiology and anatomy based on the model of the *Bildungstrieb*. The foundation of all organic structure was the force emergent from the *Zeugungssaft*. In a similar fashion more articulated structures would result from the general emergent forces that had the *Zeugungssaft* as their general basis but subsequently rendered specific through interaction with other organic or inorganic substances. Thus a hierarchy of structuring forces was conceived which, like Newton's forces of various description, had its ultimate foundation in a *Grundkraft* corresponding to a kind of Newtonian "aether" for the organic world.

An example of how Blumenbach intended to apply the *Bildungstrieb* model to anatomy and physiology is provided by his *Beschreibung und Geschichte der Knochen des menschlichen Körpers*. The structuring force of the *Bildungstrieb* was of course considered most evident in shaping the skeletal system; for as Blumenbach noted: "nature has confered upon this [part of the organism] the most powerful and active *Bildungstrieb*, since the total structure of the rest of the body depends upon it."[91] Providing the most basic structure of the organism, the skeletal system was also constructed from the most primary level of organic matter produced from the *Zeugungssaft*, to which the *Bildungstrieb* itself traced its origins. One factor indicating the primitive level of the skeletal material [*Knochensaft*] was the ease with which the organism produced it and the multiple uses it took on.

It frequently occurs that Nature uses the easily producible skeletal substance [*Knochenmasse*] to compensate for loss of the material of an organ that cannot be reproduced and makes the organ thereby at least *taliter qualiter* functional. The famous physician Morand describes a rabbit (in the *Hist. de l'Academie des Sciences de Paris,* 1770, p. 50) that had lost a foot which nature attempted to replace through a surrogate . . . by means of a foot-shaped mass of skeletal material.[92]

The base of the *Knochensaft* was an organic substance which Blumenbach called variously *Gallert* (gelatin, jelly-like substance), or *Leim.* This *Gallert* was the substance initially present in the embryo before the *Bildungstreib* began to manifest itself.

The human embryo, whose general formation does not begin until the third week after conception, consists initially of a sticky gelatin [*leimichten Gallert*]. In the following weeks . . . it attains more solidity, so that in the first half of the second month of pregnancy . . . the more solid basis of the future bones, particularly the breastbone and spinal column is clearly discernible.[93]

At this stage, according to Blumenbach, the future skeleton consists primarily of cartilage. In fact, the growth and development of bones is always preceded by the amassing of cartilaginous substance. The "more solid" gelatinous substance that makes up the cartilage was what Blumenbach called *Knochensaft.*

Blumenbach was not the originator of this theory. It had its roots in the works of Albrecht von Haller. In his *Grundriss der Physiologie* Haller had described the embryo as a sticky substance [*Leim*], which acquired solidity through the additions of earthy materials.[94] Various kinds of fibrous tissues were formed by the addition of different earthy materials to this *Leim.*[95] One such set of fibers was solidified ultimately into bone:

It appears to belong to the order of nature that the fibrous tissues are all generated out of such a *Leim.* . . . . That even the bone tissues are generated from a solidified *Leim* is seen easily from illnesses in which the hardest bones are transformed again into cartilage, flesh and finally *Gallert.*[96]

Blumenbach took over these ideas from his great predecessor at Göttingen, but in linking them up with the theory of the *Bildungstrieb* he added a dimension relative to the questions of speciation and the construction of the natural system. For Blumenbach the *Knochensaft,* formed somehow in the arteries, had a determinate constitution corresponding to the class of the organism. The *Knochensaft* was composed of a base, which as we have seen was identical to the gelatinous substance found in the embryo. In addition it contained what Blumenbach called *Knochenerde,* a substance that differed in its constituents according to the class of organism. *Knochenerde* almost universally contained calcium phosphate [*phosphorsaure Kalkerde*] and carbonic acid [*Kohlensauer*]. But in addition to these standard components Blumenbach remarked that, in addition to these compounds the bones of animals from different classes, for example, horses, oxen, chickens and cartilaginous fishes, contain according to the analyses of Foucroy and Vauquelin a considerable portion of magnesium phosphate which is completely absent in human bones. On the other hand human bones contain urea [*Harne*] which is not found in the bones of these animals.[97]

While Blumenbach did not believe that bones or organized structures of any sort could be constructed artificially from a chemical synthesis of their material constituents, nevertheless those constituents were the basis for an emergent structuring force. The force differed in its structuring tendencies according to the materials in which it was rooted, but it was not identical with them. Each organism contained a hierarchy of such forces all directed by the *Bildungstrieb* specific to that class of organism. By grasping the material bases for these forces, the relationships of their constituents among one another at different levels of organization as well as their interdependencies with the external environment, the naturalist could grasp the inner motive forces giving rise to the system of nature. Just as the followers of Newton could claim to know *how* it was that the solar system came to be organized in terms of planets sweeping out equal areas in equal times in elliptical paths around the sun by providing a *Naturgeschichte des Himmels* without ever claiming to know the cause of the force that actually generated that system, so Blumenbach claimed to be able to grasp the historical genesis of the system of organic nature through the model of the *Bildungstrieb* without claiming to know the cause of that force.

Blumenbach's ideas on constructing a Newtonian dynamic theory of organic nature, which emerged as a synthesis of ideas drawn from numerous sources but fashioned along lines proposed by Buffon and Haller, were well in progress before he first read Kant. Nevertheless Blumenbach profited greatly from Kant's writings on biology, beginning with his discussion of the distinction between races and species in 1785 and his examination of the necessity for basing biological science on a teleological framework of explanation in 1788.[98] It is not surprising that Blumenbach found Kant's approach much in common with his own developing ideas on biology, for Kant's views were independently worked out through a careful consideration of exactly the same sources, particularly Haller and Buffon, that had influenced Blumenbach.[99] Kant's thought directly influenced the mature formulation of Blumenbach's theory of the *Bildungstrieb*. In the writings of the Königsberg philosopher he foresaw the most fruitful means of making explicit certain delicate distinctions crucial to his theory and expanding it into a general theory of organic nature.

Elsewhere I have attempted to document in detail the relationship between these two men and the extent to which Blumenbach incorporated Kant's work into the mature formulation of his ideas.[100] The importance of Kant's work did not consist in proposing hypotheses or a system of organic nature for which Blumenbach attempted to provide empirical support; neither can it be argued that Blumenbach fancied himself a follower of Kant. Rather the work of the two men was mutually supportive of the same program, the program that I have called the transcendental Naturphilosophie of the Göttingen School. Although not deficient in original ideas about how to improve biology, a point to which we will return, Kant's main contribution to Blumenbach's work was in making explicit the quite extraordinary assumptions behind the model of the *Bildungstrieb*. As we have seen in our discussion of Blumenbach's work, the theory of the *Bildungstrieb* tottered precariously on the brink of accepting an out-and-out vitalism on the one hand and a complete reductionism on the other. It was difficult to see how in Blumenbach's view the formative force could be completely rooted in the constitutive materials of the generative substance, to the extent that altering the organization of these constituents would result in the production of different organisms, and still somehow be incapable of reduction pure and simple via chemical and physical laws to the constitutive material itself—how it could be both dependent on and independent of the materials constitutive of the generative substance. Blumenbach always seemed to skirt this issue by invoking a parallel to Newton's refusal to entertain a mechanical explanation for gravity. Kant explained clearly and forcefully why this was not an ad hoc strategem—how biological explanations could be both teleological and mechanical without being occult. Kant's own reason for doing this was that he had encountered difficulties in attempting to extend to the organic realm the categorical framework of the *Kritik der reinen Vernunft*, which had seemed to work perfectly for the purposes of establishing the conceptual foundations of physics. Blumenbach's conception of the *Bildungstrieb* did not resolve that difficulty; but it did permit the construction of a theory that acknowledged the special character of organic phenomena while at the same time limiting explanations in biology to mechanical explanations. Thus in the accompanying letter with the copy of his *Kritik der Urteilskraft,* which Kant sent to Blumenbach in August 1790, he wrote: "Your works have taught me a great many things. Indeed your recent unification of the two principles, namely the physico-mechanical and the teleological, which everyone had otherwise thought to be incompatible, has a very close relation to the ideas that currently occupy me but which require just the sort of factual basis that you provide."[101]

The essential problem, which necessarily requires for its solution the assumption of the *Bildungstrieb* or its equivalent, according to Kant, is that mechanical modes of explanation are by themselves inadequate to deal with the organic realm. Although even in the inorganic realm there are reciprocal effects due to the dynamic interaction of matter which result in the deflection from a norm or ideal—as for instance the departure of one body from a smooth elliptical orbit around a second body by the introduction of a third—nevertheless such phenomena are capable of being analyzed in some way as a linear combination of causes and effects, A→B→C, etc. This is not the case in the organic realm, however. Here cause and effect are so mutually interdependent that it is impossible to think of one without the other, so that instead of a linear series it is much more appropriate to think of a sort of circular series, A→B→C→A. This is a teleological mode of explanation, for it involves the notion of a "final cause." In contrast to the mechanical mode where A can exist and have its effect independently of C, in the teleological mode A causes C but is not also capable itself of existing independently of C. A is both cause and effect of C. The final cause is, logically speaking, the first cause as Aristotle might have expressed it. Because of its similarity with human intentionality or purpose, Kant calls this form of causal explanation *Zweckmäßigkeit* and the objects that exhibit such patterns, namely organic bodies, he calls *Naturzwecke,* or natural purposes:

The first principle required for the notion of an object conceived as a natural purpose is that the parts, with respect to both form and being, are only possible through their relationship to the whole [*das Ganze*] . . . . Secondly it is required that the parts bind themselves mutually into the unity of a whole in such a way that they are mutually cause and effect of one another.[102]

Now that such "natural purposes" exist is an objective fact of experience, according to Kant. Two sorts of evidence, both of which I have already discussed in connection with Blumenbach, confirm this. First, notes Kant, it is impossible, by such mechanical means as chemical combination, either empirically or theoretically to produce functional organisms.[103] Second, the evidence of generation, even in the case of misbirths, indicates that something analogous to "purpose" or final causation operates in the organic realm, for the goal of constructing a functional organism is always visible in the products of organic nature, including its unsuccessful attempts.

It might be objected that Kant (and Blumenbach) were overly hasty in asserting the impossibility of constructing organized bodies via mechanical means. In fact both Kant and Blumenbach were willing to admit this as a possibility. Kant was willing to admit—indeed he was strongly committed to the notion—that all natural products come about through natural-physical causation. Similarly, Blumenbach grounded the *Bildungstrieb* in the material constitution of the generative substance. But what Kant insisted upon is that even if nature somehow uses mechanical means in constituting organized bodies, and even if the process is capable of technical duplication, we are nevertheless incapable of understanding that constitutive act from a theoretical scientific point of view. The reason lies not in nature but in the limitations of human understanding. The problem is that human understanding is only capable of constructing scientific theories that employ the "linear" mode of causation discussed above. The types of objects that nature constructs in the organic realm, however, involve physical processes that require the teleological mode of causation. Since human reason is only capable of theoretically constructing (or reconstructing if one likes) objects that depend upon "linear" types of causal relation, the organic realm at its most fundamental constitutive level must therefore necessarily transcend the explanatory or theoretical constructive capacity of reason. Accordingly, the life sciences must rest upon a different set of assumptions, and a strategy different from that of the physical sciences must be worked out if biology is to enter upon the royal road of science.

To be sure, there is a certain analogy between the products of technology, according to Kant, and the products of nature. But there is an essential difference. Organisms can in a certain sense be viewed as similar to clockworks. Thus Kant was willing to argue that the functional organization of birds, for example—the air pockets in their bones, the shape and position of the wings and tail, etc.—can all be understood in terms of mechanical principles,[104] just as an a priori functional explanation of a clock can be given from the physical characteristics of its parts. But while in a clock each part is *arranged* with a view to its relationship to the whole, and thus satisfies the first condition to be fulfilled in a biological explanation as stated above, it is not the case—as it is in the organic realm—that each part is the *generative cause* of the other, as is required by the second condition to be fulfilled by a biological explanation according to Kant. The principles of mechanics are applicable to the analysis of functional relations, but the teleological explanations demanded by biology require an active, productive principle that transcends any form of causal (natural-physical) explanation available to human reason.

In order to understand the basis for Kant's position regarding biological explanations it is necessary to consider the argument set forth in the *Kritik der Urteilskraft*. This argument is extremely important for understanding the different biological traditions of the Romantic era, for transcendental Naturphilosophie can be considered as having accepted the position outlined by Kant, while the system of nature constructed by Romantic or metaphysical Naturphilosophie originated with the attempt to solve the problem concerning the theoretical construction of organized bodies that Kant had claimed must remain forever intractable. The special form of these Romantic theories, their employment of concepts such as polarity, unity, metamorphosis, and ideal types, as well as the structure of the system of nature constructed from them, were determined by their stand with respect to this Kantian problem and its resolution.

The task of the faculty of understanding, according to Kant, is judgment; that is it subsumes particulars given in sense experience under general concepts or rules. It can fulfill this task in two different ways: if the rule, law, or concept is already given a priori, then judgment is *determinate* [*bestimmende Urteilskraft*] ; if the particulars only are given and a general rule is sought among them, then judgment is reflective [*reflectirende Urteilskraft*].[105] In the first case the understanding is constitutive when applied to nature, while in the latter it is merely regulative; that is, in the first case it is objective, and in the second it is subjective.

These distinctions are important to bear in mind. Both are necessary conditions of experience but in different senses. In the *Kritik der reinen Vernunft* and in the *Prolegomena* Kant showed that the reason why the deductions of Newtonian physics are a priori necessary, and hence can be characterized as science, is that they are expressions of and rest upon the categories of the understanding. They are necessary because they are expressions of the formal principles constitutive of objects of experience. This explains why in physics, on the basis of certain mathematical deductions, an experiment can be constructed in which an expectation established a priori can be verified. But there are other necessary conditions for experience that are not actually constitutive of objects. According to Kant, it is necessary, for example, that we seek unity in experience, that we seek to unite as many different experiences as possible under the fewest number of principles. This requirement is subjective and regulative. It concerns the rules that must be followed in the employment of reason and the understanding. They are subjective rules that are not constitutive of objects of experience. Thus, such maxims as "Nature makes no leaps," or "Nature always follows the shortest path" do not say anything about what actually happens; "that is according to which rule the powers of the understanding play their game [*ihr Spiel wirklich treiben*] and come to an actual determination, but rather how they *ought* to go about it."[106] Such principles then are merely subjective guidelines and the results of their application cannot be accorded objective reality.

This distinction having been made, the question is whether the concept of *Naturzweck* or natural purpose, which as we have seen is necessary for interpreting our experience of organized bodies, is a concept belonging to the *bestimmende* or to the *reflectierende Urteilskraft*. From the preceding discussion we see that the solution to this problem lies in determining whether the notion of *Naturzweck* is capable of generating a priori deductive statements constitutive of experience.

In order to prepare the ground for deciding this issue, Kant considers several examples. The laws whereby organic forms grow and develop, he notes, are completely different from the mechanical laws of the inorganic realm. The matter absorbed by the growing organism is transformed into basic organic matter by a process incapable of duplication by an artificial process not involving organic substances. This organic matter is then shaped into organs in such a way that each generated part is dependent on every other part for its continued preservation: The organism is both cause and effect of itself. "To be exact, therefore, organic matter is in no way analogous to any sort of causality that we know . . . and is therefore not capable of being explicated in terms analogous to any sort of physical capacities at our disposal. . . . The concept of an object which is itself a natural purpose is therefore not a concept of the determinate faculty of judgement; it can, however, be a regulative concept of the faculty of reflective judgement."[107]

The result of these considerations is that it is not possible to offer a deductive, a priori scientific treatment of organic forms. Biology cannot reduce life to physics or explain biological organization in terms of physical principles. Rather organization must be accepted as the primary given starting point of investigation within the organic realm. In order to conduct biological research it is necessary to assume the notion of *zweckmäßig* or purposive agents as a regulative concept. These are to be interpreted analogously to the notion of rational purpose in the construction of technical devices, but it is never admissable to attribute to this regulative principle an objective existence as though there were a physical agent selecting, arranging, and determining the outcome of organic processes. At the limits of mechanical explanation in biology we must assume the presence of other types of forces following types of laws different from those of physics. These forces can never be constructed a priori from other natural forces, but they can be the object of research. Within the organic realm the various empirical regularities associated with functional organisms can be investigated. Employing the principles of technology as a regulative guide, these regularities can be united after the analogy of artificial products. Restraint must always be exercised in attributing to nature powers that are the analogs of art, of seeing nature as a divine architect, of imposing a soul on matter. We cannot know that there are natural purposive agents; that would be to make constitutive use of a regulative principle. In order to satisfy all these requirements it is necessary, therefore, to unite the teleological and mechanical frameworks as Herr Hofrat Blumenbach had done by assuming a special force, the *Bildungstrieb,* as the basis for empirical scientific investigation of the organic realm.

In all physical explanations of organic formations Herr Hofrat Blumenbach starts from matter already organized. That crude matter should have originally formed itself according to mechanical laws, that life should have sprung from the nature of what is lifeless, that matter should have been able to dispose itself into the form of a self-maintaining purposiveness—this he rightly declares to be contradictory to reason. But at the same time he leaves to natural mechanism, under this to us indispensable principle of an original organization an undeterminable and yet unmistakable element, in reference to which the faculty of matter is an organized body called a *formative force* in contrast to and yet standing under the higher guidance and direction of that merely mechanical power universally resident in matter.[108]

According to the position developed by Kant in the *Kritik der Urteilskraft*, therefore, biology as a science must have a completely different character from physics. Biology must always be an empirical science. Its first principles must ultimately be found in experience. In contrast to physics it can never be an a priori science. It must assume that certain bodies are organized and the particular form of their organization must be taken as given in experience. The origin of these original forms themselves can never be the subject of a theoretical treatment. This contrasts sharply with physics. Whereas in physics, for example, it is possible, knowing the law of attraction between all particles of matter, to deduce the shape of the earth, it is not possible, knowing the elements of organic bodies and the laws of organic chemical combination, to deduce the form and organization of plants and animals actually existing.

We can also see from this discussion the point from which the biology of the Romantics would take its origin, for they sought to construct biology as an a priori science on a par with physics. This was the program of Goethe, Oken, Schelling, and Carus. To do this they had to deny the claim basic to Kant's philosophy of biology that the human faculty of understanding cannot be constitutive of organic forms: That is, they denied that human reason is incapable of making *determinate* teleological judgments. The ground for their assertion, paradoxically, is to be found in Kant's own position. For according to the view we have just explored what we take to be organic bodies are unities artificially constructed by the understanding in order to fulfill the subjective demand of reason for unity in the realm of experience. On this theory, therefore, it is difficult, indeed impossible, to distinguish between the unity of organization belonging to a pile of stones and that belonging to a living being. Moreover, it is questionable in Kant's view that organized bodies could ever in fact be an object of experience. If reason is not provided with some faculty of making constitutive teleological judgments, how is it possible for organic bodies to be given in experience in the first place? How do we recognize them as such? The Romantic Naturphilosophen sought to answer this question by constructing a new theory of mental activity. There is such a faculty of judgment, they argued; it is the same practical faculty that makes moral judgments.

From his analysis of the teleo-mechanical framework that must underpin the life sciences, Kant went on to draw several methodological consequences. A principal feature of Kant's conception of natural science is that a mechanical explanation is always to be pursued as far as possible. In the organic realm, however, purposive [*zweckmäßige*] organization has to be assumed as given. This primitive state of organization was then to serve as the starting point for constructing a mechanical explanation. Of methodological significance, therefore, was the question of exactly how in practice the mechanical framework was to be related to the teleological framework, and secondly, at what level of investigation a primitive state of organization no longer accessible to analysis by mechanical models had to be assumed. Kant set out to answer these questions in sections 80 and 81 of the *Kritik der Urteilskraft*. These sections contain some of his most significant reflections on biology, reflections that contain in embryo the biological theory of transcendental Naturphilosophie.

One strategy would be to assume that species are the most primitive natural groups united by a common generative capacity. Indeed Kant had early on announced that: "I deduce all organization from other organized beings through reproduction."[109] Using this definition of natural species Kant had gone on to provide a mechanical model in which races were distinguished as members of the same species but adapted to different environmental circumstances. The source of this adaptive capacity was presumed to lie in the original organization of the species, in a set of *Keime* and *Anlagen* present in the generative fluid. In certain environmental circumstances particular combinations of these structures and capacities would be developed while others would remain dormant. Prolonged exposure to the same climatic conditions over many generations would cause these suppressed capacities of the original form of organization to remain permanently dormant. In the case of races, the characters affected were external, such as the structure of the epidermis, hair, nails, etc., while internal organization and the capacity to interbreed and leave fertile offspring remained unaffected.[110]

In the *Kritik der Urteilskraft* Kant expanded upon this model. Perhaps it might be possible, he mused, to find other types of organic unities containing the generative source of several related species. Such an idea had no doubt crossed the mind of every perceptive naturalist, he observed in a footnote, but only to be rejected as a fantasy of reason, since it was no more acceptable to permit the generation of one species from another than it was to permit the generation of organized beings by mechanical means from inorganic matter. But the hypothesis he was proposing was not at all of this sort; for this was a *generatio univoca* in the most general sense, insofar as organized beings would still be assumed to produce other organisms of the same type, but specifically different in some respect.[111]

The path to these more fundamental organic unities lay in comparative anatomy and physiology:

The agreement of so many species of animals in a particular common schema, which appears to be grounded not only in their skeletal structure but also in the organization of other parts, whereby a multiplicity of species may be generated by an amazing simplicity of a fundamental plan, through the suppressed development of one part and the greater articulation of another, the lengthening of now this part accompanied by the shortening of another, gives at least a glimmer of hope that the principle of mechanism, without which no science of nature is possible, may be in a position to accomplish something here.[112]

The correctness of such hypothetical unities, Kant argued, would have to be established through careful archaeological investigation of the remains of previous revolutions. Beginning from the common forms that had been provided by comparative anatomy and physiology, the archaeologist must

in accordance with all the known or probable mechanisms available to him determine the generation of that large family of creatures (for they must be conceived as such [*i.e.*, as a family] if their presumed thoroughly interconnected interrelatedness is to have a material basis).[113]

Analogous to the reconstruction of the real unity at the basis of the phenomena of races of the same species, an entity he called the *Stammrasse, Kant* was now encouraging the construction of larger common groupings of species, which he called *Stammgattungen.* Just as a common set of structures and adaptive capacities [*Anlagen*] were thought to ground the purposive organization of the species, so a similar plan of organization and common set of organs would underpin the purposive organization of several species. When exposed to varying external circumstances, including climate and, as we shall see, other organisms, this original form of organization would be capable of manifesting itself in several different but closely related ways, each being a different species of the same natural family.

There are important differences between the model proposed for identifying races belonging to the same species and that for identifying species belonging to the same family. Members of the same species can be identified with certainty verified by experiment. Any two organisms capable of interbreeding and leaving fertile progeny belong to one and the same natural species, according to Kant. The reconstruction of natural families cannot proceed by direct experiment, however. Resting on evidence of comparative anatomy, physiology, and archaeology, it is much more hypothetical in character. We are introduced here directly to one of those regulative unities that must characterize biology as an empirical science. What is important, however, is that even here the approach is empirical and capable of (limited) test.

A question that must immediately occur, particularly to anyone familiar with the modern Darwinian theory of evolution, is whether Kant means to infer from his model that the form here being discussed as the generative source of different species is an actual historical, ancestral form. The answer is unequivocally no. Such an assumption can only be consistent with a completely mechanical and reductionistic theory of organic form in Kant's view. To understand what he means it is important to recall the model of the *Stammrasse* once again. While this *zweckmäßige* organization is the source of all members of the same species, it is not itself represented in an actual historical individual. Kant strenuously denied the thesis then common among contemporary naturalists, including Blumenbach, that the various races of man are modifications of an ancestral race, which most took to be the caucasian race.[114] What Kant had in mind is a distinction much closer to that between a genotype and its phenotypical representations.[115] For he describes the *Stammrasse* as a generative stock containing all its potential adaptive variations. This is important to bear in mind when considering the generalization of the model at the level of families. Were it the case that the *Stammgattung* has an actual representative, say in the fossil record, then in passing from this individual to others of the same family new and different characters would have to be *added* to the existing stock and this addition would have to occur by means of some mechanical agency. Such an account, in short, runs strongly counter to his teleological conception of biology. According to Kant it can never be argued that an organism *acquires* its ability to adapt to its changing environment. That adaptive capacity must already be present in the organism itself, in the original purposive organization that grounds it. How that purposive organization came originally to be constructed lies forever beyond the reach of scientific treatment. What the archaeologist must presume is that the same *Stammgattung,* which is in reality a complex interrelation of organic forces potentially capable of generating numerous adaptive responses to the environment, underlies a group of forms having both current and extinct representatives. The earlier representatives will, in Kant's view, necessarily be less complex. Once he understands this regulative unity in terms of comparative anatomy and physiology, it will appear to the archaeologist that these earlier representatives have pressed together into single organisms forms that have been broken up and distributed among many organisms in later periods. Due to the increasing demands of the environment, the potential originally present in the *Stammgattung* is "unpacked," appearing as differentiated into more complex representatives. The role of archaeology is to provide an empirical test and guideline for the correctness of the hypothetical or regulative unities constructed through comparative anatomy and physiology. In any given epoch the same forces reign, giving rise in the end to the manifold of nature. The task of biology is to uncover the laws in terms of which those forces in the organic realm operate.

Rather than seeing these organic unities reconstructed by comparative anatomy as potential historical ancestors, it is more appropriate to view them as *plans of organization,* as the particular ways in which the forces constituting the organic world can be assembled into functional organs and systems of organs capable of surviving. Under different circumstances these *zweckmäßige Ordnungen* are capable of various adaptive manifestations, that is, the forces that underlie these plans are capable of assuming various expressions in achieving their effect, which is the production of a functional organism. Only under the conditions of a dynamic interpretation of form can we understand how, in Kant's view, it is possible for the fossil record to reveal an ever-increasing complexity of forms having the same generative source, while at the same time assuming that this complexity is not the result of an addition of characters:

> [The archaeologist] can let the great womb of nature, which emerges from the original chaos as a great animal, give birth first to creatures of less purposive form, those in turn to others which are better adapted to their birthplace and to their inter-relations with one another; until this womb has petrified, fossilized and limited its progeny to determinate species incapable of further modification, and this manifold of forms remains just as it emerged at the end of the operation of that fruitful formative force. But in the end, he must attribute the imposition of the original purposive organization upon each of these creatures to the Mother herself.[116]

From this passage we see that the system of nature Kant envisions is a dynamic one that runs through a cycle of birth, a fruitful period of growth and the development of the potential organic forms stored in it originally, maturity, and finally ultimate decay. From the undifferentiated potential of the entire system, governed by certain organic laws of adaptive combination that are expressed in definite organizational plans, the first primitive organisms emerge. Each of these purposive organizations has associated with it a reserve of energy. Like Blumenbach's polyps, this *Bildungskraft* can be used up in regenerating duplicates of the same organisms, or it can be partitioned out so as to produce adaptive variations on the same theme. Originally these organisms are simple and, as Kielmeyer will demonstrate for us, the simplicity of structure is compensated by the enormous fecundity of the organisms themselves. These organisms are governed by a *zweckmäßig* generative force; hence, they are capable of adapting to their physical environment as well as to the relationships that emerge with other organisms, "but only by taking up into the generative substance those materials alone which are compatible with the original, undeveloped *Anlagen* of the system."[117] The result is the alteration of the formative force, and the alteration consists in a modification of complexity in structure. Each such divergence of the *Bildungstrieb* must be compensated in some fashion, as for instance in the loss of the ability to produce numerous offspring or in the ability to regenerate lost parts. There are limits on the extent to which these forces can vary and still maintain their functional integrity, however. When this occurs, the period of growth is over and all species then in existence continue unchanged into the future. A revolution of the globe, or perhaps even a gradual but continuous change, can lead to the destruction of this system and its replacement by an entirely new set of dynamically interrelated organisms.

Polyps, Paramecia, and the Integral of Life: The Transcendental Naturphilosophie of the Göttingen School

The work of Blumenbach in the late 1770s and early 1780s wove together a number of strands of thought that had been elements in discussions on the philosophy of nature, natural history, and physiology at Göttingen from the late 1750s. The central issues in these discussions focused on aspects of Buffon's *Histoire naturelle,* which was regarded as a speculative work but nonetheless rich in ideas for the future development of science. Especially important for developments at Göttingen was Haller's work in physiology, particularly his introduction of vital forces, which he conceived to operate analogously to the forces Newton had discovered for the inorganic realm. In a grand synthesis of Haller's theory of sensibility and irritability and his views on the formation of animal tissues, together with ideas developed by others at Göttingen concerning the role of ideal types and the *total habitus* in classification, Blumenbach succeeded in setting forth a program for the construction of the natural system. For the natural history of organized nature that program was analogous to the Newtonian program for constructing the natural history of the solar system; just as the exploration of the effects of universal gravitation gave rise to natural history of the heavens, so the exploration of the laws governing the activity of the *Bildungstrieb* and its effects would produce the natural system of organized bodies.

Kant had been led to his own similar ideas on the subject independently but nonetheless through careful analysis of the same tradition of Enlightenment thought on biology, which he saw as having achieved in the work of Blumenbach theoretical foundations for the elaboration of a systematic treatment of organic nature. In the sections dealing with teleological judgment in his *Kritik der Urteilskraft,* Kant had explicated the basic assumptions of this approach to biological phenomena, the necessity of pursuing mechanical explanations in biology under the guidance of a regulative teleological framework; and he had attempted to justify those assumptions by demonstrating their consistency with the conclusions of his own earlier *Kritik der reinen Vernunft.* The significance of this step should not at all be underestimated, for however modern science and its historians may regard the contribution of philosophy to science, the fact was that in 1790 not only was Kant himself certain, but everyone else in Germany concurred, that he had effected a Copernican revolution in philosophy, and that henceforth the philosophy of nature must be consistent with, if indeed it did not take its origins from, Kant's critique of scientific knowledge. Moreover, in the conclusion of the critique of teleological judgment Kant had explored the methodological consequences of these assumptions necessary for the life sciences. There he had sketched in outline a theory of natural history that rested on comparative anatomy, physiology, and archaeology, and that led to a dynamic system of nature, to the unfolding and development of genetically interconnected forms of life, a system closely resembling that broached by Blumenbach in his works from the late 1780s and early 1790s, particularly his *Beyträge zur Naturgeschichte*. Indeed, Kant's *Kritik der Urteilskraft* was strong endorsement for the research program of the Göttingen School.

There were a number of key ideas in Blumenbach's work that became central to the development of what I have characterized above as the Göttingen program for natural history in the work of his students as well as in the works of other scientists and philosophers. First it is important to note that while Blumenbach's theory of the *Bildungstrieb* certainly was a form of vitalism, it was a vitalism remarkably different from that of Stahl, Wolff, or even Leibniz. Blumenbach did not think that discussion of a soul in matter or of peculiar vital powers was scientifically relevant. This point of view he took over from Haller. On the other hand he did not think that organization could be explained in terms of a set of material constituents alone. He argued explicitly against both views. In their stead Blumenbach adopted what is best characterized as an emergent vitalism: that is to say, the vital force was not to be conceived as separate from matter, but matter was not the source of its existence; rather it was the *organization* of matter in certain ways that gave rise to the *Bildungstrieb*. Organization was taken here as the primary given: the presence of organization could not be further explained in terms of unorganized parts. The *manner* in which it operated, the mechanisms employed for achieving the ends of organization, could be explained in mechanical terms, however. This gave rise to one of the distinctive features of the work of Blumenbach's students and it was a characteristic of the works of others exploring the ramifications of the background of ideas from which his work emerged; namely that in the philosophy of organic nature, mechanism was to be regarded as subservient to the ends of organization.

In addition to these ideas of general theoretical interest in the work of Blumenbach and Kant there were also ideas of a more particular nature affecting the mechanism of the *Bildungstrieb* that were especially compatible with ideas emerging in other areas of scientific inquiry and accordingly made the "Newtonian" program outlined in Blumenbach's works much more capable of further development at the hands of students and followers. One such idea was the notion, derived in part from Haller's theory of tissue formation, that each new level of organic force was associated with a fluid and that this fluid was composed of a *basis* plus some combination of inorganic or organic compounds. The notion that metals were composed of a basis, phlogiston, and various concentrations of some metallic principle, such as mercury, had long been current in chemical literature. The work of Priestly and Cavendish on different types of gases also exploited such a conceptual framework. More important for later developments, however, was Lavoisier's description of elements as having a basis, light or caloric, as part of their make-up. A similar conceptual structure could be found in contemporary theories of the electric fluid, particularly those espoused by Lichtenberg and Gren. Thus Blumenbach's mechanism for the *Bildungstrieb* was compatible in many respects with the conceptual structures of other areas of inquiry to which persons attracted by his work would naturally turn for further development of the theory.

The "Göttingen program" for natural history contained two implicit lines of future inquiry. In order to construct the natural system it was necessary to classify organisms on the basis of multicharacteristics. Such a system would entail a determination of the laws governing the operation of the various components of the *total habitus* and their interrelations. This aspect of the research program gave rise to the attempt to provide what Kant and others such as Georg Forster called a complete *Naturbeschreibung*. It concerned the external relations between objects. As we have also seen, a principal feature of the historical development of the tradition at Göttingen, beginning with the discussion of issues surrounding Buffon's work, was the view that the construction of the natural system depended ultimately on grasping the "inner" organization of things. In Blumenbach's scheme the *Bildungstrieb* was that internal force giving rise to the external characteristics of the organism. Consequently a second line of inquiry concerned with discovering the laws governing the activity of the *Bildungstrieb* was an integral part of the research program. The first line of inquiry led to research in various aspects of natural history while the main thrust of the second was in the direction of physiology and comparative anatomy.

An impressive list could be assembled of colleagues and students who worked on aspects of this program over the next two decades. Among the most significant were two colleagues of Blumenbach from his student days, Johann Christian Reil (1759-1813) and Samuel Thomas Sömmering. Among some of the lesser-known names associated with this school is Blumenbach's student, Christoph Girtanner (1770-1800). Although his work had no major impact on his contemporaries, Girtanner's aim was to work out details of the Göttingen program. He had been a medical student at Göttingen from 1780-1783, having worked closely with Blumenbach during that period. After numerous travels, including a lengthy stay in Edinburgh and Paris, Girtanner returned to Göttingen in 1789 where he practiced medicine and wrote on matters concerning both science and the political events surrounding the

French Revolution. Although he was closely connected with scientific developments in Göttingen, Girtanner never held a university position. In 1796 he published a work entitled *Über das Kantische Prinzip für die Naturgeschichte,* which he dedicated to Blumenbach. The purpose of the work was to propose a means for advancing the study of natural history through a synthesis of views held by Kant and Blumenbach. In particular, he attempted to explain how one might go about empirically reconstructing the *Stammgattungen* central to the Göttingen program.[118]

Another lesser-known student of Blumenbach was Joachim Brandis (1762–1845), whose family was important in Hannoverian political circles. Brandis' brother, a champion of Kantian liberalism, was a professor of law at Göttingen. In 1795 Brandis published a work entitled *Versuche über die Lebenskraft.* This book was very similar in its problem orientation to a work published in the same year by another lesser known Blumenbach student, Christian Heinrich Pfaff. Pfaff was from an illustrious family of chemists and mathematicians. In his student days he had also studied under Kielmeyer, with whom he became close friends, and he was the roommate of Georges Cuvier when all three were together at the Hohen-Karlsschule in Stuttgart. Pfaff's treatise, *Über tierische Elektrizität und Reizbarkeit* (1795), was typical of works that attempted to develop the Göttingen program in physiology, and it merits our attention.

The question motivating Pfaff's treatise was whether all vital functions are modifications of a single *Grundkraft* or whether they are specific individual forces. Moreover he hoped to determine whether the vital forces were different in kind from mechanical forces. This question could be solved, he supposed, once the general form of the laws of organic force could be determined and the attempt had been made to reduce them to the general laws of physics.

In this treatise Pfaff focused his investigation on the vital forces of sensibility and irritability. Since the phenomena of animal electricity were manifestations of these two vital forces, a determination of the laws regulating galvanic phenomena would lead to a general force law for sensibility and irritability.

Since the phenomena of animal electricity are phenomena connected with the sensible and irritable parts; and since as phenomena connected with life, they presuppose vital forces as their first causes, we must direct our attention to these forces first before we seek the principle which sets them all in motion. The phenomena of animal electricity can be seen as signs or revelations of the relationships between certain external circumstances and those two forces. . . . [t]hus we must concern ourselves first only with the phenomena of sensibility and irritability, and in particular mainly with the forces upon which these depend, since the phenomena of animal electricity are merely manifestations of these forces. And we want especially to strive to express the interrelationship of these two forces which are so closely linked together.[119]

In pursuing these goals Pfaff conducted a great number of experiments establishing the conditions under which contractions could be excited in a prepared frog's leg by contact with metal conductors, and the relative strength of the contractions for various substances. From the generalizations derived from these experiments, he attempted to draw an analogy between these phenomena and the phenomena connected with electricity; his aim being to argue that the vital principle underlying sensibility and irritability was in fact the electric fluid, and that the forces of sensibility and irritability were "positive" and "negative" manifestations of this unitary _Grundkraft_.

Pfaff's work on animal electricity was extremely influential. The experiments and reflections recorded in the treatise served as the starting point for the researches of both Alexander von Humboldt and Johann Wilhelm Ritter. And in demonstrating that the phenomena connected with animal electricity were fundamentally chemical in nature, Ritter's work first established in a convincing way that the unity of organic and inorganic forms in nature postulated by Pfaff was not purely speculative.

For the later development of the Göttingen program and for German biology as a whole in the nineteenth century the most important figures were Heinrich Friedrich Link, Carl Friedrich Kielmeyer, Gottfried Reinhold Treviranus, and Alexander von Humboldt. Each of these men was a student of Blumenbach and they all maintained close contact with him over the years. It is in their work that we find the transcendental Naturphilosophie of the Göttingen School worked out in systematic detail.

Within this group the most impressive contributor by far to the Göttingen program—one whose contributions spanned detailed painstaking empirical research in organic chemistry, comparative anatomy, and physiology (particularly galvanic phenomena, plant and invertebrate physiology), as well as deep and powerful thoughts concerning the theory of the natural system—was Carl Friedrich Kielmeyer. Having studied previously at the Hohen-Karlsschule in Stuttgart, Kielmeyer moved on to Göttingen where he studied with Blumenbach, Gmelin, and Lichtenberg from 1786 to 1788. He returned to the Karlsschule from 1790 to 1793, during which time he lectured on comparative zoology as well as chemistry and natural history. He returned once again to Göttingen for several months during 1794. Kielmeyer was thus a participant in and, as we shall see, a lively contributor to the intense discussions on the construction of a theory of animal form going on in Blumenbach's circle during the late 1780s and early 1790s.

In his lectures at the Karlsschule Kielmeyer assembled into a grand and comprehensive program the various aspects of the approach to constructing a general theory of animal organization that I have sketched from the writings of Blumenbach and Kant. Although these lectures were never published, their contents were widely known, and copies of the lectures must have circulated. In a letter to Windischmann, Kielmeyer mentions that copies of these manuscripts were circulated. Cuvier's correspondence with Kielmeyer's student, Christian Heinrich Pfaff, demonstrate that while Cuvier did not receive copies of Kielmeyer's manuscripts he was following the development of Kielmeyer's thought in these lectures.[120] References to these lectures in the writings of Döllinger, von Baer and others leave little doubt that they must have been widely known.

In addition to stating the conditions for a materialistic interpretation of the teleological-mechanical conception of the phenomena of organization I have sketched from the works of Kant and Blumenbach, and stating the implications for generalizing the model for constructing a natural system, Kielmeyer's lectures made an essential contribution by describing a path for beginning to implement these ideas given the existing state of biological and chemical science. Two essential problems demanded solution. First, although ultimately the proposed scheme required that the basis of each type of organism lay in the system of organic chemical affinities embedded in the first instance in the generative substance, the analysis of organic materials had only just begun; and although the French chemists in particular had made some advances in this area, still no satisfactory application of chemical methods to the general theory of animal organization could be expected in the forseeable future. Kielmeyer, who made extensive and substantial contributions to the development of *Pflanzenchemie,* the beginnings of organic chemistry, was deeply sensitive to this problem.[121]

The second problem concerned the actual construction of the natural system viewed as a genealogical system based on the laws of generation and reproduction. As Blumenbach had noted, even though the natural system must be based on generation as a theoretical principle, the practical application of the breeding criterion is circumscribed within certain definite limits.[122] Although different races of the same species are theoretically capable of interbreeding, slight differences in periods of fecundity and differences in behavioral characteristics might set up natural barriers to interbreeding even among members of the same species.[123] Moreover, the breeding criterion was obviously useless for higher taxonomic levels. Blumenbach proposed as the solution to this problem the use of multiple characters in classifying organisms: based on comparative anatomical and physiological investigations animals were to be grouped together in accordance with their agreement in *total* number of characters. Kielmeyer built upon this idea.

In his lectures on comparative zoology Kielmeyer set forth a plan for constructing what he called the *Physik des Tierreichs.* Its design was to develop methods for revealing the laws of organic form through comparative anatomical studies of mammals, birds, amphibians, fish, insects, and worms. The program consisted of a multifaceted investigation of animal organization, first through a comparative study of the chemical basis or organization. This was to be followed by a comparative anatomy and physiology of basic organs as they exist fully developed and in the various periods of embryonic development. Here attention was devoted to three groups of organs. First those concerning the relation of the organism with its *external* environment; namely digestive organs, the lymphatic system, circulatory system, the brain and nervous system; also included in this group was a comparative study of sensory organs and the investigation of systems of motion; namely muscles, bones, and their "analogues" in various animal forms. The second group of organs for study were those concerned with the regulation of the *internal* functions. Here Kielmeyer included comparative studies of the kidneys and the various other "regulatory" glands of the animal economy. The third and final group of organs to be considered were those that served for the *communication* of the animal with other members of its species, namely organs of generation. Kielmeyer also included in this group the comparative anatomy and physiology of organs of speech, *Stimmorgane.*[124]

After establishing the "elements" of structure in the organic realm Kielmeyer's program proceeded to a general theory of the relations between them or to an *Allgemeine Physiologie der Tiere.* Here Kielmeyer advocated the use of developmental histories of the genesis of the germ and its material constituents, the subsequent development of the embryo, and finally the development of the mature organism and the changes it undergoes in relation to its environment. Since the principles regulating each type of organic form lay locked up in the *Keime* and *Anlagen* of its generative substance, comparative developmental histories would reveal interrelations between different organic systems; nature itself would provide, so to speak, its own experimental laboratory. By systematizing and unifying the patterns through which form is unfolded more general relations would emerge from which general laws could be constructed.

Thus far Kielmeyer had presented the methods for revealing the laws of the "deep structure," the internal forms of organization. In turning to an analysis of the external surface elements of form Kielmeyer attached special significance to behavioral studies as a means of understanding the principles of organization. He advocated the construction of a *Psychologie der Tiere.* Its object was to study a) the activities in terms of which animals seek out nourishment, a proper climate, and suitable habitat; b) activities through which they defend their position in the economy of nature against enemies.[125] Animal psychology was also to include the investigation of activities that promote the preservation of the species, among which he included mating behavior and the rearing of offspring.[126]

Like Blumenbach and Reil, Kielmeyer believed that a systematic study of the variation to which animal forms are subject and the patterns of these anomalies would provide positive insight into the principles of organization. Consequently, he advocated the construction of a *vergleichende Pathologie der Tiere* as a third methodological tool to be employed in the new science of zoology. Here "permanent, inborn as well as accidental variations of species would be investigated; and chiefly under two classes of variation, 1) malformations, monstrous births, bastards; variations with respect to geographical location and other (similar) circumstances; inheritable degenerations and permanent, inborn variations induced by climatic and geographic variation; universality of variation; b) variations in capabilities of the organs and their stimulation; temperament, both individual natural temperament and characteristic idiosyncracies."[127]

Kielmeyer summarized the various aspects of his *Physik des Tierreichs* and the order of their application as follows:

a) The number of organs in the machine of the animal kingdom or the number of animal forms generally and the laws according to which these are divided into different groups. Causes, consequences, or purposes [*Zwecke*].

b) The relative position of the organs in the machine of the animal kingdom, or the division of the animal kingdom into groups upon the earth (geography) according to different characters. Laws of the differences according to different groups. Causes and effects.

c) The interrelated formation of organs in the animal kingdom. Gradation of animals and affinities in their formation generally as well as according to groups. Laws, causes, and effects of this gradation.

In the next category Kielmeyer introduced an area of study which he had not previously mentioned in the outline of his lectures, namely paleontological research:

d) Changes the animal kingdom and its groups have suffered on the earth. *The developmental history of the animal kingdom* in relation to the epochs of the earth and those probable for our solar system. Symbolized by the parabola.

e) Changes the animal kingdom and its groups undergo repeatedly [throughout all epochs]. The life of the machine of the animal kingdom or its *physiology*. Symbolized by the circle.[128]

In a concluding section of this manuscript, which Kielmeyer crossed out, the *Physik des Tierreichs* was characterized generally as a kind of Laplacian dynamics of animal organization according to which the series of animals and the elements of their organization were to be viewed as a series of attempts by nature to break up the integral of life into a series of partial fractions.

From this plan of a general science of animal form sketched in his lectures

we see that Kielmeyer, in addition to uniting the various elements characteristic of the approach of Blumenbach and the Göttingen School, had begun to
introduce a completely new dimension to the discussion, namely the use of
the embryological criterion for detecting affinities between animal forms. To
be sure this was to some extent implicit in the earlier notion of a generative
stock shared by different groups of organisms and the related interest in inheritable degenerations and malformations, but the idea of utilizing embryogenesis as a means for investigating the unity of the generative stock was
Kielmeyer's most significant contribution.

Kielmeyer expanded upon his notion of the biogenetic law in a treatise,
which like almost all of Kielmeyer's work was never published. It was written
in 1793-1794 and entitled "Ideen zu einer allgemeinen Geschichte und
Theorie der Entwickelungserscheinungen der Organizationen." Several aspects
of Kielmeyer's conception of the relationship between phylogeny and ontogeny presented in this manuscript provide an important context for later
developments in Germany.

Kielmeyer begins by pointing to a fundamental difference between the results to be expected from teratology and embryology. Malformations appear
to be dependent on external circumstances, such as environment, and while
they are probably rooted in the matter of the germ, they are departures from
the rule and are not repeated similarly in all individuals. Embryological development, however, always reveals a patterned series of successive changes
that is the same for each individual of the same species and patterned differently for different species.[129] These patterns of embryogenesis are, therefore,
more dependent on an internal directive force: they tell us more about the
internal organizing principles of animals, which as we have seen depend not so
essentially on the chemical conditions of life as much more on the *order* and
arrangement of those conditions. For Kielmeyer the beauty of focusing on
embryological patterns was that "they demonstrate the path and contents of
the system of animal organization as a whole without requiring the assumption of a special directive force existing outside of the individual organism,
through which the life and economy of organic nature is maintained."[130]
That is, recourse need not be taken to a *Weltseele*, to any supra-material
organizing force. Furthermore, although in his view embryological investigation is the most useful means for constructing a general theory of animal
organization, it can also aid in the construction of natural classification,
which most "descriptive" biologists regard as the highest aim of their science

insofar as the relationships between the different forces and different forms
of manifestation of the same force in different organisms is exactly that
which determines the essence of the differences and relationships between
species. With the determination of these forces, therefore, and the laws they

obey, the path toward constructing the natural system would be given at the same time.[131]

In a letter to Windischmann of 1804, Kielmeyer explained the reasoning behind his postulation of an interdependence of the results of embryological and paleontological research in his earlier lectures at the Karlsschule.

The idea of a close relationship between the developmental history of the earth and the series of organized bodies, in which each can be used interchangeably to illuminate the other, appears to me to be worthy of praise. The reason is this: Because I consider the force by means of which the *series* of organized forms has been brought forth on the earth to be in its essence and the laws of its manisfestation *identical* with the force by means of which the series of developmental stages in each *individual* are produced, which are *similar* to those in the series of organized bodies. . . . These forms, however, demonstrate a certain regular graduation in structure as well as similarity to the stages of individual development; therefore it can be concluded that the developmental history of the earth and that of the series of organized bodies *are related to one another exactly* and therefore their histories must be bound together.[132]

Kielmeyer went on to add an extremely important qualification to this thesis. He wanted to emphasize that in his view this "series" of forms must not be conceived as *continuous*. There are gaps in the developmental series that can never be filled, not simply because of defects in the fossil record, but because there are different types of organization.[133] Like Blumenbach, Kielmeyer denied the existence of a chain of beings.[134]

Nevertheless, while Kielmeyer denied the existence of a continuous developmental series, he did argue for the transformation of species and the interconnection of forms within the intervals punctuated by the gaps in the developmental series.

Many species have apparently emerged from other species, just as now the butterfly emerges from the caterpillar. . . . *They were originally developmental states and only later achieved the rank of independent species;* they are transformed developmental stages. Others on the other hand are original children of the earth. Perhaps, however, all of these primitive ancestors have died out.[135]

He went on to note that, like Lamarck—and though he is not cited in this context, Blumenbach—he believed that the production of these genetically related but distinct forms "was due to an altered direction of the formative force introduced by changes in the earth."[136] But this alternation of the *Bildungstrieb* did not proceed continuously. In Kielmeyer's view the "paths through which the different series of organisms has been brought forth have been very different in different periods of the history of the earth."[137] Thus, not only were the genetic relations between groups of organisms to be viewed as circumscribed within definite limits due to the internal organization of different types, but the manner in which these fundamental organizational plans were worked out in different periods and the (limited) developmental series of organisms descendent from them were dependent upon and circumscribed by the external conditions prevailing within a given geological age.[138]

We might summarize the general theory of natural history emerging from Kielmeyer's works as follows. There are definite epochs of nature, during which a different flora and fauna, specific to that epoch, flourish. Within each of these epochs the same laws regulating animal organization prevail, just as the same laws continue to regulate inorganic phenomena. Each epoch contains a system of interrelated organisms based on a small number of ground plans. Within each epoch gradual transitions occur within the forces of both the inorganic and organic realm. As gradual shifts in environmental circumstances occur within an epoch the *Bildungstrieb* of the primitive forms are modified, giving rise to divergent phylogenetic lines of organisms within the same type. Although the forms of the next epoch are based on the same principal plans there is no continuation of the previous forms. A change in one element of the system entails a modification in all the others, for each individual form is related to the whole of organized nature. Each epoch, therefore, is its own complete, closed system; and it is not possible to trace a single phylogenetic line, even within the same ground plan, from the most recent epoch.

The quintessence of the position developed in his unpublished lecture notes was distilled elegantly by Kielmeyer in his famous lecture delivered at the Karlsschule on February 11, 1793, entitled, "Über die Verhältniße der organischen Kräfte untereinander in der Reihe der verschiedenen Organizationen: Die Gesetze und Folgen dieser Verhältniße." This paper, approximately forty pages in length, is one of the milestones of the Romantic era; anyone wishing to understand the biology of this period would do well to examine it carefully.

The lecture begins by discussing the general methodological framework that must be assumed if success is to be achieved in constructing the system of nature. The framework is that set forth by Kant in the *Kritik der Urteilskraft:* the constitutive causes of organic nature cannot be grasped. Nature must be treated as if it employed a technique analogous to purposive action, one that relates means to ends in teleological fashion. The definition of an organized body, following Kant, is one in which all its parts are reciprocally cause and effect of one another.[139] In a literary vein, but one reflecting Kant's powerful imagery of the great womb of nature as well as indicating that the most fundamental secrets of nature can at best be reflected in a story conscious of its analogy to purposive human activity, Kielmeyer himself speaks forth as *die*

Natur. To underscore the necessity but at the same time the futility of ever penetrating the secrets of organization through teleological judgments, Nature is asked what her intentions were in constructing this multiplicity of forms. Her answer is: "I had no intentions, even though the intermingling of cause and effect appears analogous to the connections your reason makes between means and ends; but you will find it easier to understand these matters if you assume such a linkage of cause and effect as though it were in reality one of means to ends."[140]

Lyonet and Bonnet had estimated at least seven million different organic forms on the surface of the earth. Each of these is represented by at least 10,000 different individuals. Each individual in turn is constructed from as many as 1,000 to 10,000 organs. In order to make a system out of this fullness of life, according to Kielmeyer, it is necessary to understand the forces that are united in and generative of these individuals. Next it is important to understand the relationship of these forces with respect to one another in different species of animals and the laws according to which this relationship changes in the series of organic forms. "Finally the task is to understand how both the continuity and change in species are grounded in the causes and effects of these forces."[141]

In answer to the first question—what are the forces united in individuals?— Kielmeyer identifies five forces: sensibility, irritability, reproductive power, power of secretion, and power of propulsion. In order to measure these forces and compare them to one another, he proposes that the strength of vital force be conceived as a compound function of a) the frequency of its effect, b) the diversity of this effect (i.e., the number of diverse forms in which it is manifested), and c) the magnitude of the opposition it encounters from other forces. In the absence of an exact measure and until one satisfying the demands of this function can be constructed, Kielmeyer notes that in essence a vital force is one that demonstrates "permanence of effects under otherwise constant conditions,"[142] a definition that seeks to identify vital force as the source of regulative maintenance of the organized body. The similarity in formulation to Newton's principle of inertia—the force of inactivity as it was then understood—is strong.

Kielmeyer's plan in the work was to look at each of the five vital forces considered singly, and then compare each of their strengths within different species of animals. Beginning his examination with sensibility, Kielmeyer notes that the capacity for retaining a diversity of types of sensations specifically different from one another falls off in a graduated series beginning with man. In the mammals, birds, snakes, and fish all the same sense organs as in man are present, but the degree of complexity of these organs differs for the different classes and even within the same class. In the insects the organ for hearing is absent, while the sensitivity to odors is much enhanced; and even if the eye appears multiplied a thousandfold in these animals, it is for the most part immobile and only capable of admitting light in a few species. In the worms, finally, all the diverse organs of the other species are replaced by a single sensibility to touch and light. It must not be overlooked, Kielmeyer tells us, that when in the series of organic forms one sense organ is lost, hence diminishing the diversity of the effect of the force of sensibility—component b) of the function above—greater opportunity for the development of one of the other senses is afforded; and when one sense is less developed, another will be more sensitive, its organ more delicately structured.

From these observations we derive the following law: The diversity of possible sensations falls off in the series of organic forms in proportion to the increase in the fineness and discrimination of the remaining senses within a limited domain.[143]

A little reflection revealed that this law is not exactly correct, that even within the same class of animals the reduction in capacity of one of the senses is not always compensated by an increase in another. The ground for departure from this first law Kielmeyer sought to find in the law governing the effects of the second force named above, namely irritability. In contrast to sensibility, irritability manifests variations not only in the diversity of its effect [component a) in the definition of vital force], but also in the frequency of its manifestation in a given time and in the length of its manifestation under similar circumstances [i.e., components b) and c) from the definition].

In the mammals and in the birds, if the trunk is severed from the head, and individual members from the trunk, all traces of irritability vanish within a short time. Cold-blooded animals exhibit quite a contrary set of phenomena. Frogs can hop around with their heads removed, and decapitated turtles can move around with their hearts removed for several days.[144] Kielmeyer noted similar observations for spiders and fish.

The phenomena lead to the conclusion that irritability *increases* its strength and independence from the rest of the organic system in the series of organisms beginning with man. Looking toward other characteristics associated with this phenomenon, Kielmeyer notes that most of the animals that tenaciously preserve this power of irritability are animals in which either very few irritable organs are present or ones in which the muscles are separated from one another. Mussels, for instance, which exhibit a high degree of irritability, have at most two or three distinct muscles.[145] Fish, while possessing numerous muscles, have only a small number of different types of muscles, in contrast with man, where there are relatively few muscles but a great variety of muscle-types and complexity. Moreover, those animals capable of preserving irritability in the highest degree are also those that move the slowest. From all of these observations Kielmeyer derives the following law:

Irritability increases in the permanence of its manifestation in the same proportion as the speed, frequency, or diversity of its effect and as the multiplicity of different types of sensation decreases.[146]

The second law, therefore, provides the needed corrective factor to the first law, for we see that in the series of different organic forms deficiency in sensibility is compensated by an increase in irritability. But it provides only part of the needed correction, Kielmeyer tells us. The force of irritability cannot be preserved as long in mussels, or even in plants, as it can in amphibians. Another force must be sought that affects irritability, accounting for its departure from the norm in certain forms.

Kielmeyer finds the needed modification in the force of reproduction. As a first approximation to the law of the reproductive force, he notes that the mammals normally produce one to fifteen offspring, while birds produce many more than fifteen, and some species of amphibians produce at least one hundred thousand. Examining these phenomena more closely, Kielmeyer observes that the animals that bring forth the fewest offspring in each class are those having the largest bodies. Thus rats give birth to from ten to fifteen offspring at once, while whales produce only one calf. Furthermore, it appears that the less prolific animals are also those having more complex structure and the ones whose offspring require the most time to come to term. "Thus it takes nature two years to make an elephant, while only a few weeks suffice for constructing a rat."[147] These observations result in the following law for the reproductive force:

The more the reproductive force is expressed in the number of new individuals, the smaller are the bodies of these new individuals, the less complex are they, the smaller is the period required for their production, and the shorter is the active period of this force itself.[148]

As in his discussion of the previous laws, Kielmeyer went on to point out several exceptions to this one. The exceptions in this case, however, were only apparent. Thus, while some insects are less prolific than certain fish, it is exactly these insects that exhibit the greatest number of metamorphoses or possess the capacity for regeneration in the greatest degree. Similarly the least prolific amphibians, namely the lizards and snakes, are also the ones capable of achieving the largest body size. Also, according to Kielmeyer, the least prolific mammals and birds are exactly the ones that exhibit the greatest degree of difference in their sexual organs; species of insects and worms exhibiting unlimited growth and high capacity for regenerating damaged parts are also the ones in which sexual differentiation is either absent or in which both sexes are very similar. Kielmeyer was, however, willing to acknowledge certain exceptions to the operation of the law of the reproductive force, but he thought they could be clarified by determining the influence of the external medium in which the animal lives and also the effect of temperature on the reproductive force.[149] These considerations led finally to a reformulation of the law of the reproductive force:

The more so we find all the different modes of reproductive force united in a single organism, the sooner do we find sensibility excluded, and the sooner also does even irritability disappear.[150]

Having made a comparative study of the three most important forces in his original list, Kielmeyer turned to a consideration of their relations with respect to one another. The system implied by his preceding analysis is obviously a dynamic one. Taken together his three laws imply that in the series of organic forms, sensibility is gradually superceded by the reproductive force. Irritability too is finally superceded by the reproductive force, the increase in one of these forces being compensated by a decrease in one of the others. These are the *internal* forces giving rise to animal form and function, and while they do not operate independently of external forces such as the medium, temperature, etc., they are the only sources of animal structure. These forces alone, the same forces operating in every individual, give rise to the entire structure of the organic realm. This point, as we have seen, was essential to the Göttingen program, and it was especially emphasized by Kant: a purposive unity of forces must give rise to the organic realm. The same forces must operate at all levels of differentiation bringing forth families, species, races, varieties, and ultimately individuals. The individual carries in it the organic forces that differentiate it as a member of each of these higher collective unities. This differentiation cannot at all come about as a result of accidental external modification of inorganic nature. Rather the conditions for bringing forth specifically different types of organisms must always lie in those organisms themselves, in the purposive interrelation of the organic forces productive of organic bodies. External factors provide the conditions for expressing now one permissable expression of these forces and then another, but the true source of this manifold diversity lies in the internal forces of organization.

Fundamental to Kielmeyer's conception, therefore, a point he emphasized at the beginning of his lecture, is that the same set of forces united in every individual, though expressed in different degrees, are also the forces that give rise to the entire system of organic nature. This led to the major claim of the paper, and to Kielmeyer's greatest contribution to the Göttingen program; namely that the order in the appearance of these forces in the generation of an individual is the same as the order of appearance of these forces in the system of nature. Ontogeny recapitulates phylogeny:

The simplicity of these laws becomes evident, when one considers that the laws according to which the organic forces are distributed among the different forms of life are exactly the same laws according to which these forces are distributed amongst individuals of the same species and even within the same individuals in different developmental stages: even men and birds are plant-like in their earliest stages of development; the reproductive force is highly excited in them during this period; at a later period the irritable element emerges in the moist substance in which they live—according to experiments which I have made on chickens, geese, and ducks, even the heart is possessed of almost indestructable irritability during this period—and only later does one sense organ after another emerge appearing almost exactly in the order of their appearance from the lowest to the highest in the series of organized beings, and what previously was irritability develops in the end into the power of understanding, or at least into its immediate material organ.[151]

This principle—that the distribution of forces in the series of organized beings is the same as the division between different developmental states of the same individual—offers a means for constructing the system of nature. According to it the lowest classes are the ones in which the reproductive force is most pronounced. These we might call *Reproductivtieren.* Being characterized by a prolific reproductive and regenerative capacity, this class will contain among all other classes the greatest number of species. Included in this class will be the worms and insects. Similarly there will be *Irritabilitätstieren* and *Sensibili-tätstieren,* these classes corresponding to the invertebrates, amphibians, mammals, and birds. Within these various classes of animals the same pattern will be repeated; animals possessing the greatest reproductive power will stand first (or lowest) and so forth.

An important aspect of Kielmeyer's theory is that in neither the lecture on the series of organic forces nor in any of his other lecture materials did he ever assert that the series of beings is linear, so that the ontogeny of man recapitulates the phylogeny of the entire animal kingdom. Although he never explicitly developed the system in detail, the evidence of his writings seems to suggest that he regarded each class of animal as having various interconnected sets of organs as the material expression of the system of forces grounding them. In the *Sensibilitätstiere,* for example, the sense organs were the predominant organizing principle of the class, although it is clear that being the highest class, all the organs of the other classes must also be available to them. These animals would then specialize in the development of one or more of the sense organs. There could be varying degrees of development of each of these organs. Due to the dynamic interrelation of all the organic forces, the particular preponderance of one (or more) sense organs would entail a corresponding functional arrangement as compensation among the other organs. The system resulting from this scheme would not be linear but rather radial in structure. At the core of the stem for each group must be imagined not an actual animal, but the specific purposive combination of organic forces (the five named above) containing *in potentia* all the organs and combinations of organs that will be developed by the different species of the group. Different species will correspond to the developmental grades of this primary functional unity. The series of forms developed will not be such that each developmental grade of a particular organ or closely related system of organs follows upon one another in a tight temporal series. Much more consonant with Kielmeyer's view that all animal forms limit one another is the notion that several different species of the same family develop simultaneously, each one representing a developmental grade specialized on a different organ system. Viewed in this manner, Kielmeyer's system is quite compatible with that sketched by Kant in the *Kritik der Urteilskraft*, but in it one can see rudimentary traces of ideas that would be developed more clearly and systematically later on by Karl Ernst von Baer.

At the end of his lecture of 1793 Kielmeyer entertained the idea that all of the organic forces he had discussed were in fact different manifestations of a single unitary force, a *Grundkraft*. Perhaps, he said, the entire machine of nature derives its motion from a single force originally awakened into action by light.[152] This was an idea similar to that proposed earlier by his teacher Georg Christoph Lichtenberg in a short paper written in 1778, entitled, "Über eine neue Methode, die Natur und die Bewegung der elektrischen Materie zu erforschen." But it was also an idea that would have a strong resonance in the mind of the young Schelling, who in nearby Tübingen was dreaming of the *Weltseele*.

Although Kielmeyer's lectures attempted to set forth a systematic theoretical account of the program designed to explore the physics of organization, he did not publish the results of any empirical research that actually implemented the program. There were others, however, who shared Kielmeyer's views, but who did attempt to realize the aims of the program through extensive empirical research. One such individual was Alexander von Humboldt.

Not only are all the theoretical and metaphysical components of what I have been describing as a Newtonian research program persistent themes in Humboldt's writings, but of all those who worked in this tradition Humboldt came closest to realizing its goals. The traces of Blumenbach's formative influence on Humboldt's research plans are evident in his first publication, *Aphorismi ex doctrina physiologiae chemicae plantarum* (1793). The physiological investigations that Humboldt set forth in his treatise were intended to reveal the laws governing the activity of the *Bildungstrieb*. The definition of vital force in terms of which the investigation proceeded bore unmistakable marks of Blumenbach's influence.

I call *Lebenskraft* that internal force which dissolves the bonds of chemical affinity and hinders the free interaction of the elements in the body. Thus there is no more certain sign of death than the decomposition through which the elements [*Urstoffe*] re-establish their previous rights and order themselves according to the laws of chemical affinity.[153]

Like Blumenbach, he pointed out that any theory of organic form required some special vital organizing principle. There was sound empirical evidence for assuming the existence of such an organizing force: Humboldt pointed out that of the thirty-seven known elements, only eighteen were to be found in organic bodies. More important, however, was the fact that these elements were found in combinations that human art was incapable of reproducing.[154] It appeared therefore that some vital principle was capable of overcoming the forces of chemical affinity and combining elements into specific organic forms.

Although Humboldt's principal objective in this treatise was to draw together the results of his chemical researches on plant physiology, indirect evidence in the work indicates that he envisioned the research as contributing to the construction of the natural system. The treatise, which appeared originally in Latin, was translated into German in 1794 at the bidding of Johann Hedwig, to whom Humboldt had dedicated the work. Hedwig had also worked on plant physiology. In the foreword composed by the German translator, we find an interesting linkage between Humboldt's physiological researches and the goal of constructing the natural system.

Although Linnaeus was in actuality more than a mere systematist, . . . his work has had a disadvantageous effect on the development of natural history [*Naturgeschichte*], in that for more than half a century, natural history has been investigated only with respect to problems of classification.

As a result the careful description of nature [*die Naturbeschreibung*] has been neglected and accordingly numerous aspects remain without application, which would otherwise open the prospect for connections between the various branches of science.[155]

Since Humboldt was consulted on the translation of his work from the Latin it is likely that the ideas expressed in the foreword reflected his own view of its importance. Independently of that point, however, it is clear that Hedwig, Christian Friedrich Ludwig, and Gotthelf Fisher, who collaborated on putting out the German edition, saw the work as providing the kind of research necessary for advancing the study of natural history. By investigating the effects of various gases, and the effects of light, electricity, and magnetism on plant physiology, Humboldt was providing a data base from which descriptive regularities regarding forms and functions of plant life could be derived. Through this effort a deeper insight into the *total habitus* of plant life would be obtained from which a natural classification would ultimately emerge.

Humboldt's commitment to the Newtonian research program extended into areas other than physiology. He was also concerned directly with problems of classification. This particular interest is evident in the early work, but it is most conspicuous in treatises composed after his expedition to South America. The *Ideen zu einer Physiognomik der Gewächse* in particular bears evidence of the concern for constructing an ideal typology characteristic of the tradition at Göttingen that we have been exploring. Humboldt writes in the *Ideen,* for example:

In spite of a certain freedom in the abnormal development of individual parts, the deepest [*urtiefste*] force of organization binds all animal and vegetable forms to fixed, eternally recurring types.[156]

Through comparative studies in the botanical gardens of Europe as well as through observations made during his expeditions, Humboldt had come to the conclusion that all genera and species of plants could be reduced to a small number of ideal types.

If one comprehends in a single glance the different phanerogamous varieties of plants which are already housed in herbaria and whose number is currently estimated to be more than 80,000, one recognizes in this amazing number certain primary forms to which most of the others can be reduced. . . . Sixteen such plant forms determine primarily the physiognomy of nature.[157]

Like Blumenbach, Humboldt regarded these ideal types as internal forces giving rise to the basic structure of organisms. It was his aim to find the laws in terms of which external factors diverted the activity of these internal forces, giving rise to the various classes of organized beings found in nature.

In the enormous cats of Africa and America, in the tiger, lion and jaguar, the form of one of our smallest domestic animals is repeated in a larger measure. If we penetrate the inner crust of the earth . . . we find a distribution of forms which not only no longer agrees with those of the present climatic zones: they also reveal colossal forms which hardly differ at all from those presently existing [except in size alone]. If the temperature of the earth has undergone considerable, perhaps periodically recurrent changes, if the relationships between sea and land, or even if the height of the atmosphere and its pressure has not always been constant; then the physiognomy of nature, the size and structure of organization itself, must have already been subjected to numerous changes.[158]

Just as in the works of Kielmeyer, one sees in Humboldt's writings the intention of constructing a dynamics of organic form. What he attempted was to construct mathematical regularities relating change in form with change in temperature, geographical distribution, and geological change.

Humboldt considered this work as partially accomplishing some of these ambitious ends, particularly in regard to the variables affecting plant distribution. He proposed a statistical method for studying plant distribution, a method that agreed with that proposed by Alfonse Candolle and Robert Brown, although he claimed to have hit upon the idea independently.[159] What he did was to find for any given zone, such as that between the fifty-fifth and sixtieth north parallels, the ratio between the number of natural families and the total number of phanerogamous species.[160] An alternative approach consisted in simply comparing the ratios of the absolute number of varieties of species belonging to each natural family in a given zone; but for the purpose of drawing general laws, Humboldt preferred the method of forming ratios with respect to the total number of phanerogamous species.

Humboldt's method for constructing the system of nature had two distinct quantitative components. On the one hand it involved the construction of an ideal typology based on comparative anatomical studies. Independent of that, however, was the statistical study of the distribution of plant forms and the numerical relations between genera and species in different geographical and thermal zones. Humboldt contrasted these two methods of investigation and the questions to which they gave rise in the following terms:

The quantitative relationships of plant forms and the laws that are observed in their geographical distribution can be considered from two distinct aspects. If one considers the arrangement of plant forms according to natural families independently of their geographical distribution, one asks: what are the fundamental forms, the types of organization, that lie at the basis of the formation of the classes? What is the relationship between the monocotyledons and the dicotolydens? These are all questions of general phytology, the science which investigates the organization of plant life and the relationships between organisms presently existing.

On the other hand, if one considers the classes of plants which have been united in terms of similarities in structure not according to the abstract method but rather in accordance with their distribution over the earth, a different set of questions emerges. One investigates then, what plant families are more dominant in the torrid zone than toward the polar circle. . . . Do the forms which cease to dominate in moving from the equator toward the poles follow the same law of decline as the pattern of dominance in the ascent of equatorial mountains?[161]

As we have seen, both of these methods, both the use of an ideal typology and the determination of the laws governing the *habitus,* were integral parts of the Göttingen program for natural history. Although Humboldt added new elements to Blumenbach's original formulation of the program, particularly in his use of statistical methods for studying plant distribution, these were not fundamental modifications of the research program as envisioned by Blumenbach and reflected in the works of Kielmeyer. There can be no doubt that Humboldt was deeply influenced by Candolle, LaPlace, Werner, Forster, and others, but what they provided were insights into means for bringing about a realization of the Newtonian research program for natural history that he had encountered as a student in Göttingen. This point emerges most clearly perhaps in the following passages from Humboldt's last and greatest work, *Kosmos,* in which he discussed the necessity of unifying the two methods mentioned above:

The systematically ordered register of all organic forms, which used to be designated splendidly by the term "natural system," offers an amazing linkage of form (structure) in accordance with internal principles. . . . not a linkage according to spatial grouping, viz. according to geographic zones, altitude, influences of temperature, etc., which affect the entire surface of the planet. But the highest goal of the physical observation of the earth [*physische Erdbetrachtung*] is . . . to grasp unity in multiplicity, to undertake research into the inner connection of terrestrial phenomena. Where particulars are mentioned [in this science] it is only in order to bring the laws of organic arrangement into agreement with the laws regulating geographical distribution. . . . The natural series of plant and animal structure is thus something given, as taken over from descriptive botany and zoology. The task of physical geography is to trace how quite different sorts of forms, although apparently dispersed randomly over the surface of the earth, stand nonetheless in a secret genetic relationship to one another [*in geheimnissvoller genetischer Beziehung zu einander stehen*].[162]

The science Humboldt sought to establish and which he called *physicalische Weltbeschreibung* was, like Kielmeyer's comparative "world" zoology, a dynamics of organized nature; it was a causal account of the interconnection between the structures in "static" taxonomic systems:

In grasping nature, Being is not to be distinguished absolutely from Becoming. . . . In this sense *Naturgeschichte* and *Naturbeschreibung* are not to be treated separately. The geologist cannot grasp the present without the past. Both interpenetrate and coalesce into one another . . . just as in the field of languages, the entire past process of language formation is reflected in the present. In the material world, however, this reflection [of the past] is even more apparent. . . . *Its form is its history.*[163]

In reflecting on the implications of this dynamic approach to organic form, Humboldt was led to the same conclusion, explicitly stated in the works of both Blumenbach and Kielmeyer: to wit, that the imbalances in the forces of nature have led to the destruction of whole groups of organisms and their replacement by others,[164] and that simple forms of the *Urwelt* have been divided up, "dissected" as it were, and spread out through time by the forces of the *Nachwelt.*[165]

In a footnote to his *Rede* of 1793 Kielmeyer mentioned that he was at work on a general theory of the organic realm in which the dynamic system he had outlined would be worked out in all its empirical details. The work never appeared. Had it been completed it would have undoubtedly borne strong resemblance to the six-volume work published by Gottfried Reinhold Treviranus entitled *Biologie: Oder Philosophie der lebenden Natur* (Göttingen, 1802-1822). This work was the crowning achievement of the transcendental Naturphilosophie developed by the Göttingen School. Gathering together the best empirical research of the day, Treviranus attempted to set forth the natural system in bold panorama. The structure of that system, the assumptions upon which it rested and the general view of the organic realm it espoused differed in almost no detail from that set forth by Kielmeyer in his magnificent lectures on the "Physik des Tierreichs."

The stated object of Treviranus' *Biologie* was to construct the natural system: to determine the conditions and laws under which the different forms and phenomena of life exist and their causes. The science that treats these matters was henceforth to be called "biology, or the theory of life."[166]

In order to set out upon the construction of this new science Treviranus argued that agreement must first be reached on a definition of life. Life must, according to him, be viewed as a structured *internal* activity giving rise to form, growth, and motion. While the source of this activity is an internal cause, it expresses itself only in relation to external phenomena; and accordingly every motion that it brings forth is necessarily a mechanical one.[167] The question arose, then: how is this activity to be distinguished from mere external, mechanical motion? After rejecting several proposals, among them those of Stahl and Alexander von Humboldt, which he regarded as too vitalistic, he settled upon a definition of life as consisting in the capacity to produce a continuity and an apparent necessary interconnection among phenomena while reacting to accidental influences originating in the external world, a definition close in content to that proposed by Kielmeyer in his *Rede.*

Treviranus grounded this concept of life in Kant's theory of matter. All matter must be thought of as the result of a momentary equilibrium among opposing forces. Kant had argued in his *Metaphysische Anfangsgründe der Naturlehre* that at the basis of the concept of matter must be thought the unity of an attractive and repulsive force. This approach made all matter self-limiting, the source of its structural boundaries being generated from within. Treviranus made an important modification of this Kantian theory, which had significant consequences for his understanding of the generation of organic form. He argued, by contrast with Kant, that a single type of *Grundkraft,* namely repulsive force, is sufficient. If one assumed a manifold of such independent centers of activity, the result would be the mutual self-limitation of these forces, the momentary equilibration of which could account for the same material phenomena Kant had sought to explain with two forces. Treviranus attributed this improvement in the Kantian theory to

Schelling, but Schelling was wrong, he argued, in postulating a single hyper-physical *Grundkraft* as the unity of these opposing forces: "Force" is that which stands at the limit of our capacity to inquire further into the appear-ances of material nature. Furthermore, "force" must be thought as something finite, according to both Kant and Treviranus. It is not possible to think of a force without also conceiving another force opposed to it. Therefore, while two opposing forces may indeed be united, it could not be a third, unopposed force that unites them. This third thing must, therefore, lie beyond the limits of a possible physical account. Schelling's *Weltseele* was, therefore, a meta-physical hypothesis having no place in natural science.[168]

These reflections on the construction of matter led Treviranus to two important conclusions. First that no change in the world is possible without disrupting the equilibrium of forces in some small neighborhood. With every expansion of one force must follow the contraction of another and vice versa; every chemical change must produce a mechanical change; similarly, mechan-ical rearrangements must lead to changes in all other forces such as electricity, magnetism, and light.[169] Every material system must, therefore, pass through an infinite series of changes, without ever returning exactly to the same point from which it started:

The series of changes through which every material system passes must be so constructed that after several revolutions it returns nearly to a point where it had been previously without, however, ever returning to it exactly. Each material system is best represented therefore in the form of a spiral.[170]

We see here the theoretical foundations being laid for establishing a claim to be made later on that certain material systems, namely major groups of animals, including species, have been transformed into other genetically related forms as a result of continuously changing conditions in the external world. Also implicit in this position was the notion that since the system of material conditions never returns exactly to the same point, it is impossible to reconstruct through experiment the conditions that obtained when the present forms of life were being generated.

The second result of Treviranus' reflections on the construction of matter concerned the conception of organized matter directly. Since all the mani-festations of the *Grundkräfte* were tightly interconnected, each was simulta-neously cause and effect of the other, means and at the same time end. Nature, including inorganic nature, had, therefore, to be viewed as a single, unlimited organism.[171] The only difference between the inorganic and the organic realms was in the degree to which this mutual interdependence of cause and effect is evident. In the organic realm, moreover, the number, order, and interconnection of forces is much more finely tuned. Inorganic matter, according to Treviranus, is characterized by *actions,* and these can be exercised in relative independence of one another. The life of organized bodies on the other hand is determined by functions, i.e., closely interconnected actions such that each is at the same time cause and effect of the other. Because of this functional interdependence of forces in the organic realm, organic bodies are much less adaptable to changed conditions; the equilibrium of forces is much more easily disturbed. There are accordingly certain definite boundaries set by the functional interdependence of these forces beyond which it is impossible to go without destroying the life of the organism.[172] Moreover, in keeping with the force model set forth in the beginning, Treviranus assumed that, as functional unities of forces, every organism must be opposed by other limiting forms of life:[173]

These forms are not otherwise intelligible than under the assumption that the different classes and orders of organized beings differ not only in degree, but also in their mode of receptivity for external influences and are capable of opposing these external influences with reactions of their own. This difference in receptivity and this reactive capacity, however, can only have its ground in a difference in form of organization. Accordingly, there are as many different types of organization as there are different forms of life. And to every form of life there must correspond a particular type of organization.[174]

Treviranus went on to argue that within the same type of organization the same principle must apply. There must be a mutual limitation of forms leading to a differentiation and gradation of different modes of the fundamental organizational unity of the class. In discussing Kant and Kielmeyer I have argued that a central notion of the Göttingen program was that all the different groups of related organisms in the natural system had to be conceived as related through the development of the adaptive potential of an original purposively organized ground plan. In these opening arguments of his *Biologie* Treviranus was seeking to provide a theoretical grounding for this model in his teleological reconstruction of matter.

Treviranus' conception of the *Lebenskraft* and the *Bildungstrieb* also places him in the direct line of ancestors of the Göttingen School. There are three ways of conceiving organic forces, he argued. According to the first system the vital force is a direct product of inorganic materials. This view was rejected, for it implied the possibility of an artificial production of organic substances, which Treviranus regarded as impossible. Furthermore, it seemed to him that for many inorganic elements, such as carbon and calcium, nature is dependent ultimately on organic bodies.

The second system held that the vital force is simply superimposed on inorganic materials, directing them into a purposive organization. This position was likewise swiftly rejected. If one assumed it, commitment was necessarily made to the view that organisms can under no circumstances bring forth forms different from themselves. According to this view, every organic form must only reproduce its own kind. Such a view could not explain, therefore, the source of interconnection among large families of existing forms, nor could it explain the apparent interconnection of forms throughout the history of the earth.

Since all matter must be conceived as organized in different degrees, according to Treviranus, it followed that there must be a sort of ground state of organization, a *vita minima*, from which all other higher forms of organic matter emerge.[175] Like Haller and Blumenbach, he assumed the existence of a basic organic *Grundstoff*. Treviranus himself related this notion directly to the work of Needham and Buffon, a point to which we will return later.

Organic matter must in itself be formless but capable of receiving every form of life. By combining with inorganic materials it takes on a special form and structure. Moreover, this form of organization must be different according to the difference in these inorganic materials.[176]

Once again Treviranus was laying foundations in terms of which he could deal with the phenomena deemed central to the theory of organic nature by the Göttingen School. On the one hand organic materials could be dissolved into the materials of the inorganic realm, and physiological processes could be understood in terms of chemical and physical mechanisms. On the other hand, it was not possible through chemical and mechanical means to produce organic material artificially; some organic substance must always be present to act as a catalyst. In one fell swoop Treviranus proposed a materialistic model in which the teleological and mechanical frameworks were neatly linked. This accomplished, he attempted next to ground this conception of organic matter in empirical data and to extend it to the construction of the natural system.

In constructing the natural system two principles must be followed, according to Treviranus. From the preceding discussion we have seen that each type of organization in the organic realm demands its own particular *Mischung* or specific set of chemical constituents, which it joins to the *Grundstoff* of the organic realm; its own *Textur*, which Treviranus took to be the forms in which these chemical elements are united into organs; and *Struktur*, the overall anatomical arrangement of the organic parts and the specific external identifying characteristics. These latter two categories depended ultimately on the first, according to Treviranus, and hence the ultimate goal in biology was to press on in understanding the operations of organic chemistry well enough to be able to generate the structure and texture of organic bodies completely from their material constituents alone.[177] Such a science of biology would never be attainable by human reason, however. It would be an objective and constitutive science, which, as we have seen from our analysis of the Kantian foundations of the Göttingen program, was deemed unobtainable for us.[178] The science of biology must remain for us a teleological discipline: "In the classification of living organisms the rule must be followed: wherever the chemical mixtures are well known, to deduce the chief characteristics from these; but where these are not sufficiently well known, to take recourse first to the texture and only as a last resort to structure as a means of understanding the dependent characteristics."[179] This implied that it is not the general shape of the animal and the most striking structural resemblances that must be taken as the focal point of classification, but rather it is the *organs,* the internal elements of structure that must be taken as primary. This followed from the dynamic theory we have seen outlined both by Kielmeyer and by Treviranus in the opening sections of the *Biologie.* According to this theory, material systems were distinguished in degrees of complication by the number of points of contact they have with the external world (i.e. their sense organs and nervous system), by their ability to react to the impressions they receive (i.e. the organs of motion and respiration), and by their generative capacity (i.e. reproductive organs and digestive system). Related organisms specialize in differentiating in one or more directions these related systems of forces. Hence focusing on overall shape and similarity of structure rather than upon gradations in the relationship of these forms can result in the oversight of extremely important connections in the natural system.

Recognition of the teleological character of biology implied another extremely important principle. Unlike physics, it was impossible for the biologist to start with simple elements and their laws of interaction and deduce from these laws the shape and texture of nature. Not only because of the complex nature of organic beings, but also due to the different mode of causation employed in the organic world and the specific structure of human understanding, which is not suited to deal directly with that sort of causation, biological organization must be assumed as *given.* In our case, this turns out not to be a hindrance for constructing the natural system. We are capable of grasping the natural system because *we are living, organized beings.* This point cannot be discussed fully here, but it is so singularly important to the entire biology of the Romantic era that it should be noted with some care.[180] Since we are forced by the limited nature of our understanding to interpret nature by constructing a framework analogous to the sort of causation employed in human practical, technical decisions, it is necessary that the biological given we take as the model for the framework be *man himself.* In moments of scientific frivolity, Treviranus tells us, we might pretend that biology is a science that starts from the *Ursuppe* and builds organized bodies from the least to the most complex without reference to man; but in point of fact, for all the reasons we have discussed above in reference to Kant, this is simply

impossible.[181] Man is and must remain for biology the model on which the
natural system is to be built. As it turns out this is not a source for regret and
condolence, because man belongs to the class of organisms, namely the
vertebrates, that contains the most complicated organisms in the system of
nature, and within that class man is the most complex organism. This is not
a subjective judgment, it is an objective fact confirmed every day by experience. Man is the highest, most complex, and fully developed organism in
nature. That might change, and indeed Kielmeyer, Treviranus, and the entire
Göttingen School thought it would—that a revolution of the globe might
bring forth a new set of organized beings. But the necessity of falling back on
man as the model for the present system suffices. According to Treviranus,
however, this could not be the case if man occupied a different position in
the system. How do we know that the present system will not bring forth
more complex animals? Kielmeyer provided the grounds for an answer. The
productive period of our epoch is over; new species are no longer being generated. Man is currently in possession of the fullness of his being, capable of
developing all his capacities. It is the Age of Enlightenment, the age when
men are capable for the first time of acting in terms of ethical judgments
that are the practical fulfillment of the principles of reason.

   Having set forth the principles for constructing the system of nature within
the teleological framework, Treviranus set to work on organizing the animal
kingdom. Following Cuvier,[182] he divided the animal kingdom in two classes:
red-blooded animals with an internal articulated skeleton, and white-blooded
animals with either no skeleton, an articulated internal skeleton, or an external skeleton. It is instructive of the approach to biology being developed here
to follow the construction of the first class, the red-blooded animals, within
which Treviranus included the mammals, birds, amphibians, and fish.

   Treviranus began by naming the organs all the animals in this class have in
common. In addition to the characters already named, they all have a brain
protected by an enclosed skull. "The brain is always divided in two halves;
they all have a double optic lobe; a cerebellum; four ventricles, including a
double frontal ventricle and two unpaired ventricles. . . . one almost always
finds at least three types of sense organs, sight, hearing, and smell."[183] After
inventorying the parts of the vertebrate brain, Treviranus went on to list the
parts of the eyes and ears that all these animals have in common.

The ears [for example] always have three semicircular bony or cartilaginous
canals which likewise contain the same number of membraneous semicircular
canals. These swell at the places where the acoustic nerves enter forming a
sac-shaped hollow, which is always surrounded by and filled with labyrinth
fluid. . . . The muscles in these animals are united by cellular tissue; the heart
always lies beneath the brain and above the digestive and generative organs
between the respiratory organs. The heart can have one or two chambers. . . .

The maxillary bone of these animals always lies horizontal, and always opens downward, by which motion a tongue is revealed. . . . The digestive tract always stretches from the mouth. All have a peritoneum which encloses the digestive organs and all have a liver. . . . most also have a spleen and a pancreas. . . . All have two kidneys, distinct sexes with two gonads in the male sex.[184]

These organs and the general plan of their organization formed the core of the vertebrate class. It is important to emphasize that for Treviranus, as for the others in the Göttingen School, this group of organs, partially listed above, was not a mere descriptive catalogue. As we have seen from the dynamic system discussed by Kielmeyer, these organs and their interrelations were thought to be the material expression of a group of closely interdependent vital forces. The purposive or *zweckmäßig* character of these forces led them to be expressed as organs and organ systems. These were the *zweckmäßig Keime* and *Analagen* [seeds=organs=Textur, and organ systems] upon which Kant had argued the dynamic theory of the natural system must rest. Accordingly, each of the different orders of the vertebrate class was envisioned as a functional differentiation of this fundamental unity. The different species of each family of animals was viewed as a different grade of this functional type, and so forth. In order to reconstruct the path of this process of differentiation and thereby the order of families, genera, and species within each group, it was imperative that the most fully developed, structurally complex species be taken as the model for the whole family: Thus for the mammals, man was the prototype:

We can regard man as the prototype in regard to the formation of the mammals. . . . It can be assumed that the shape of the human body can be transformed into that of the other mammals through the lengthening or shortening of the different parts. Thus the difference between the human skull and that of the other mammals consists only in that the latter are more oblate, and that the line drawn from the base of the nose to the foremost incisors (or the area where these teeth are located in man) is almost perpendicular to the plane in which the lower end of the teeth of the upper maxillary are found, while in the other mammals it makes a larger or smaller angle with this plane. Thus the angle is greatest in the elephants and the apes, least in the deer, dolphin and the anteaters, where it lies almost in the plane itself.[185]

Treviranus went on to note that, excluding the cervical ligament and tail, man has all the anatomical components of the other mammals, but each different group has developed some of these characters to a higher degree of specialization. The human brain, on the other hand, has many special characteristics lacking in the other animals of this class.

After completing a comparison among the various mammalian organ systems, Treviranus proceeded to determine the order among the various families of mammals according to their degree of divergence from man.[186] As a focal point indicative of other connected divergencies among organ systems, Treviranus chose the hand as the key to his ordering of the mammals. Thus the apes stood next to man as a separate family, then came animals with claws, namely the canines, the bats, which were the connecting link to the next family, the rodents, and finally the sloths. Each of these classes was also characterized by divergence from or complete lack of some characteristics found in man. Thus, in consideration of their pinnate members, the whales came next, their teeth and multiple stomachs placing them close to the sloths and rodents.

The hoofed animals followed. There were three different orders in this subclass: *Porci,* or *Schweine; Pecora* or *Rinde;* and *Equi,* or *Pferde,* these divisions being made according to the number of toes or clefts on the hoof. First in this class were the *Schweine,* or pigs, among which Treviranus included the rhinocerous, tapirs, elephants, hippopotomae, and the hogs. This was an extremely difficult class to organize, according to Treviranus. It has numerous anomolous genera closely linked neither to other genera within the order nor to genera of other orders. Such anomalies led Treviranus to remark that these forms almost appear as if they could be the remains [*Überbleibsel*] of forms that flourished in an earlier age of the earth and that were destroyed by a revolution of the globe.[187]

This was a significant observation, for it underscored a major difference between the biology of the Göttingen School and the evolutionary theory later to be developed by Darwin. The system envisioned here was conceived to be a causal theory of the natural system; and because it investigated the laws productive of animal form as the condition for establishing the natural system, it was regarded by its proponents as a *generative* theory. Now it was assumed that at the basis of each class lies a group of organs and particular modes for arranging them according to the same plan. The order in which these forms appear in the fossil record, however, need not reflect the natural order among individual species, but only of major functional groupings. It is extremely important to bear in mind that, for this theory, species were not the natural units upon which nature works. Species characters were the most external adaptive modification of the vital internal functional unity of forces lying at the basis of a major group. Species were part of the fine tuning that this functional unity made in adapting to external conditions. The purposive organization at the basis of any class of animals was assumed to provide the condition for its adaptation to a range of habitats, but the Göttingen biologists did not regard the prior existence of any one of those specific adaptive responses to be a necessary precondition for the development of another. This would have resulted in the total dependence of form upon the environment, a position they persistently denied. If the natural sequence of potential developmental grades of a stem were completely determined by the sequence of environments, and not by a set of organic laws providing the internal source of form and function, then there would be no need to assume a purposive organization in the first place, and certainly no need to ground classification on the laws of biological organization. It is not surprising, therefore, that Treviranus did not identify the *Porci* [*Schweine*] as the species *from which* the other higher mammals evolved, even though they together with the horses and oxen occupy the lowest developmental rank among the mammals in his scheme. For he did not argue that these species, which may indeed be the remnants of forms from a previous period of the earth, are in fact the physical ancestors of these higher forms.

A similar approach was followed by Treviranus in ordering the birds. He selected the ostrich as the archetype of this class. The reason for this choice, was that the ostrich is the closest link between the birds and the mammals, being most nearly connected, according to him, to the camel:

It has more hair than feathers over most of its body; the long neck bends in the same manner as that of the camel; its thighs are thick and not muscular; its sinewy feet have only two toes, just like the camel; its wings are more like arms; its upper eyelids are mobile just as in the mammals, and it is provided with lashes, just as in man and the elephants. . . . the male has a penis which is much longer than that of the other birds and which is very similar to that of the mammals, and the female has a kind of clitoris. The ostrich forms the transition therefore, between the last two families of the mammals and the birds.[188]

When turning to the amphibians, however, Treviranus employed a different approach. In this class he did not select a single archetypical form to serve as the model for the rest of the class. Nor did he select the feet and toes as a clue to other relations. Instead he pointed to gradations in the complication of three different but ultimately closely interdependent organ systems: the heart and respiratory system, the acoustic organs, and the reproductive organs. The acoustic apparatus proceeds in a seriated gradation, he observed, from the turtles, where it possesses nearly the same degree of complexity as it has in the birds, to the frogs, where it is similar in structure and complication to the corresponding organ in the fish. A similar series is revealed with respect to the structure of the heart. This series, according to Treviranus, is the most significant with respect to the overall organization of the class, for "the differences in the structure of the acoustic organs and the reproductive organs as well as all the other organs and the reproductive system are parallel to the differences in the structure of the heart."[189] These considerations led him to two different orders of amphibians; namely those with either a double atrium and three ventricles or those with only a single atrium and ventricle.

Differences in several other characters led Treviranus to divide the first order once again into three classes, so that taken together the amphibians contained four orders; the turtles, lizards, snakes, and frogs.

After rounding out his ordering of the vertebrates with the classification of the fish, Treviranus went on to apply the principles of his new science of biology to arranging the invertebrates and plants. At the conclusion of these efforts, he re-emphasized the main points of his approach in addressing the question of whether there is a chain of beings from man to the infusorians and mosses. The answer, he said, is both yes and no. If single organs of a single organ system are considered in isolation, then nature offers the appearance of a single continuous chain; but there are many organs and hence "a thousand and even many thousands of chains which are woven together with infinite artistry into the tightest knots to constitute the whole of nature."[190] These "knots" are functional unities constituting a *zweckmäßige Grundform* at the basis of each class:

> In every family, in every genus, even in every species of living being nature develops some organ or system of organs principally, while she leaves some of the others unchanged and still others more simplified; and this articulation as well as the related simplification are normally the repetition of the same plan. [*Grundform*].[191]

Central to this whole conception of biology is the dynamic postulate stated by Kant and Blumenbach and developed by Kielmeyer: nature never increases the complication of any organ or organ system without diminishing in compensatory fashion the complexity of some other related organs. The source of dynamic and purposively organized forces lies within the organism itself. In order to understand the manner in which these internal principles of organization are expressed fully, however, it is necessary to consider their connection to the opposing forces of the external world in response to which they are capable of generating adaptive modifications. Treviranus turned to this theme in the second volume of the *Biologie*.

In discussing the relationship of organization to the external world, Treviranus began from a postulate which he went on to ground with a mountain of evidence, namely that the characteristics of classes, families, genera, and even species stand inseparably connected to the organization of the environment in which they live.[192] "A preference for this or that element is often the only factor wherein many families, genera, and species are distinguished."[193] Moreover, it is extremely important to note that the characteristics Treviranus had in mind were not simply morphological characters. He also regarded behavior and temperament as characteristics bearing upon the place of an animal in the economy of nature.[194]

The first half of his treatment consisted in a factual overview of what was known about the geographical distribution of plants, zoophytes, and animals. Several generalizations emerge repeatedly from his data. First he was struck by the dynamic interdependence of climate, habitat, and the forms of life that occupy them as well as the interdependence of these forms upon one another. Recorded history provided much evidence indicating that the local economy of nature has been changed by man. Through the cutting of forests, damming of streams and rivers, draining of bottom lands, etc, many animals had been displaced, some becoming nearly extinct and being replaced by others over the same geographical range. Man himself had, he noted, directly altered many animals through domestication. But man was not alone in possessing this power. What he has accomplished mirrors what nature herself has done through natural and gradual physical causation. Changes in climate, geological forces, and the relation of animals to one another result in parallel phenomena. But one rule that seemed to prevail among all these variations was that whenever the same external conditions are present, the same types and gradation of animals appear. Whether in the Alps or the Andes, wherever we find the same temperature, soil constitution, and general atmospheric conditions, the same kinds of animals are present. Whatever slight variations found in otherwise similar conditions were due ultimately to physical conditions, which could never be duplicated exactly.[195]

In discussing the internal organization of plants and animals above, we have seen that central to Treviranus' conception was the notion that the overall system of forms in each major group can be treated as a radial system branching out from a central stem to several knots, which form the center for a bundle of rays, which in turn branch out in all directions. Treviranus saw the relationship of organisms to their external environment as reflected in the distribution of forms as exactly parallel to this internal structure. The plant kingdom, for instance, "can be compared to a tree whose trunk is in the polar regions and whose branches are extended southwards over the earth, in that as they approach the limits of the southern temperate zones, they multiply and separate ever more from one another. In their first emergence many of these branches form smaller side-branches through which anastomoses are formed once again with the trunk."[196] The same proposition held for the animal kingdom: "It can be maintained with certainty that all these animals form a tree, just as in the plant kingdom, in which the multiplicity of genuses and species increases in a seriated fashion [*Stufenfolge*], which is only interrupted by local circumstances from the polar circles to the equator."[197] These two trees of plants and animals were not independent of one another, however; they stood related in a most wondrous harmony: "All land mammals, all birds and most of the amphibians, fish and insects are linked in their distribution almost completely to the distribution of the plants."[198]

A final important proposition related to Treviranus' use of the tree metaphor and its connection to the physical distribution of plants and animals was his conclusion that there are certain geographical centers for each genus, in the neighborhood of which the greatest number of species of that genus are to be found. Thus for the dicotyledons, there were eight such geographical centers.[199]

For the science of biology, as Treviranus defined it, it was not sufficient that a set of empirical generalizations be drawn from the data of natural history. Natural history must not only have its Kepler, it must also have its Newton. Biology was not *Naturbeschreibung;* its aim was to seek explanations, or following the usage most consistent with Kant's teleology, its aim was to provide *Erörterungen.* Ultimately it must seek to trace the grounds for these empirical generalizations back to the postulates of the dynamic theory of matter under the guidance of the principle of teleology. In the third section of his treatn ent of the distribution of plants and animals, therefore, Treviranus turned to the construction of a theoretical model for explaining these phenomena.

The principal thesis of Treviranus' work was that all living organisms are the product of physical forces, that the same forces that have produced different forms of life in different epochs are identical with those still in operation, and that the effect of these forces differs only in degree or direction as a result of external conditions.[200] This being the case it was required to seek some set of phenomena among the present activity of these forces which could provide a clue to understanding the gradation of forms and their interrelation. Fortunately, the solution lay immediately at hand. It was provided by the experiments on vegetable matter and infusorians of Priestly, Needham, Buffon, Wrisberg, Müller, Ingenhouss, and (in spite of himself) Spallanzani.

The details of the controversies surrounding Needham's infusorians are well enough known not to require discussion here.[201] Some of the highlights that found such strong resonance within the Göttingen School and in the work of Treviranus in particular are worthy of attention, however. The discussion of Needham's experiments always occupied a central position in Blumenbach's works, particularly in his work *Über den Bildungstrieb.* It is probably not insignificant that Kielmeyer was promoted to his position at the University of Tübingen with a treatise entitled "Observationes quaedam ad investigandum ortum animalculorum infusionum."

Needham's conviction that spontaneous generation could not be effected from an artificial synthesis of inorganic materials in the absence of organic matter or an organizing principle, such as heat or light, was well suited to Treviranus' purposes.[202] This principle was further confirmed by experiments done by Wrisberg. In Priestly's vegetable substance, the *grüne Materie,* Treviranus saw that formless organic matter capable of generating all forms in connection with the external influence of the environment that was so central to his whole conception of organic nature.

Almost paradigmatic for his explanation of the relation between form and environment were Needham's infusions with wheat kernels. The kernels gradually degraded into a kind of gelatin [*Gallert*] with numerous fibres. After a while these began to move, transforming into veritable plant-like zoophytes. In the next stage small movable particles of a different shape were seen to emerge from this first generation. These latter ceased to move after a few days. They then united into a large mass from which new spherical shaped zoophytes emerged.[203] Herein Treviranus saw a beautiful confirmation of his definition of life as a phenomenon expressed by the mutual interaction of vital and physical forces; for only from this definition, he had argued, could it be possible to understand how one species of animal can give rise not only to members of its own kind, but transform into other different forms when altered conditions no longer favor its preservation.

Wrisberg's observations gave even more exciting evidence of such transformations. In his "Observationes de animalculis infusioriis satura" Wrisberg had observed that vegetable or animal infusions containing neither acids nor anything that would hinder fermentation would, shortly after the first appearance of air-bubbles, contain a multitude of tiny circular objects that would after a while become enclosed by a thin membrane. These tiny molecules were the building blocks of all plants and animals. They differed in no major appearance from the smallest infusorians, except in their inability to move, and the lowest grade of infusorians emerge directly from them. In infusions where these molecules did not develop, infusorians did not emerge either. Once one of these tiny molecules had become mobile, it united with others to form a larger animal. These in turn united to form larger infusorians. Often a small section would separate from one of these and move off independently. Wrisberg observed that the presence of these larger organisms always followed upon the presence of the smaller, less complex ones. Equally important from Treviranus' perspective, Wrisberg had noted that his populations of infusorians limited one another and replaced one another in cyclical fashion. In an infusion of fly larvae he observed first the molecules, then formations of small animals of very simple structure; then came some with a fish-like appearance, others oval in shape, and still others he identified as polyps. The fish-shaped infusoria and polyps underwent a period of decline that was directly associated with a decline in the population of the smallest infusoria. After these latter had all but disappeared, the whole process started up again, first with an increase in the smallest infusoria, followed by the fish-shaped ones, and finally by the polyps.[204]

The importance of these observations for the foundation of his theory led Treviranus to set up experiments of his own with infusoria, which he reported at great length. He found the observations of Needham and Wrisberg fully confirmed by his own. He too saw a successive series of ever more complex infusoria being constructed out of an original set of spherical molecules. In one experiment he saw a single large spherical-shaped infusorian divide in two, both parts of which continued to move independently. Likewise he saw the interdependence of forms and alternate contraction and expansion of populations of paramecia, rotifers, and vorticells. In one experiment the "epoch of the paramecia lasted only about ten days."[205]

All of these considerations on infusorians emboldened Treviranus to make a major conceptual leap. The major thesis of his earlier investigation of the internal principles of organization had been that each new division of animals consists in an addition in the kinds of organs constituting the animal accompanied by a corresponding increase in potential complexity and in the number of contacts with the external world.[206] As we have just seen, Treviranus thought this same pattern could be seen being repeated directly on a small scale in the experiments with infusoria. The tiny molecules were the first "organs"; combinations of these led to a differentiation in kind and complexity in a succession of organisms:

Supported by the analogy with the zoophytes, plants and lower classes of animals, therefore, we may assume that the *Urformen* of the mammals and birds were once generated in that same manner in which now only the zoophytes are still formed.[207]

The justification for making this analogical leap of faith rested on the assumption of the continuity in the operation of natural forces. In countering the objection that nothing directly comparable to the construction of such complex forms as birds and mammals is observable in the present operation of these forces, Treviranus called upon the dynamic theory of the epochs of the earth developed by Kielmeyer:

It cannot be objected that a similar generation of higher classes ought to be observable if this theory were true. For what occurred in that period when the system of nature was coming into being can never happen again once it has already achieved its organization.[208]

In concluding this discussion Treviranus drew attention to the fact that the theory of organic nature he had developed in the second volume was not original with him. It was really the theory first proposed by Buffon and Needham, but which they were unable to develop consistently. Like his own theory, theirs rested upon two postulates: first, that a single organic substance is the material basis of the entire organic realm, and that this matter is capable of receiving any form; second, that nature has certain formal principles, *innerliche Formen* or *moules intérieurs* by means of which this organic matter is structured from within. But each of these forms could only retain its particular structure so long as external influences remained constant.

Through continuous action of gradually and only slightly modified circumstances, these organisms could take on a new form. This new form was, so to
speak, the result of the new equilibrium established between the internal system of forces constituting the animal and those forces in the external environment.[209]

In the last few pages of the second volume, Treviranus provided an overview
of the essential external influences that affect the form and distribution of
plants and animals. The empirical evidence for these factors was explored in
great detail in the third volume of the *Biologie*. They were the same factors
to which Blumenbach had first called attention in his small treatise on the
Bildungstreib and in his book, *Beyträge zur Naturgeschichte:* slight changes in
the chemical constitution of the soil, air, water, etc., can ultimately effect
changes in the sources of nutrition and eventually in the constitution of the
generative fluids; when sufficient numbers of individuals are thus affected
over long periods of time, a modification of form can result. The reason for
these changes was that inorganic matter could enter into chemical combination with the basic *Urstoff* of organic matter, generating thereby the material
potential for structurally more complex forms of organization. The specific
manner in which these forms were organized depended upon a) the internal
organization of the organic (vital) forces already present in the organism, the
internal forces that account for generation, growth, and nutrition, and b) the
forces of the external environment that either permit these internal forces
free reign to develop in a particular direction or throw up some hindrance to
such a development. The model for this theory, we have seen, was provided by
Blumenbach's *Knochenlehre* and by Haller's work on tissue formation. Finally, among the material external factors of change mentioned above, Treviranus
placed great emphasis on another that both Kielmeyer and Humboldt thought
to be extremely important: temperature. An empirical generalization he drew
from his investigations on this subject was that "the multiplicity of types,
number and the size of organic beings stands in direct proportion to the
gradation of temperature."[210]

From all of these postulates and the empirical data he had assembled to
support them, Treviranus drew two conclusions that express in capsule form
the philosophy of nature of the Göttingen School:

From these postulates the original products of organic nature can be explained. With the emergence of these first products, however, new forces were
awakened which influenced the formation of the following generations. Foremost among these [forces], however, is to be included the dynamic effect
which every organism has on the rest of nature.[211]

We believe that encrinites, pentacrinites and zoophytes of the prehistoric
world are the original forms from which all the organisms of the higher classes
have come into being through gradual evolution [*Entwicklung*]. . . . And it

appears to us to follow that, contrary to what is commonly said, the animals of the prehistoric world were not destroyed by great catastrophes; rather many of these forms have survived, but they have disappeared from nature because the species to which they belong have been transformed into other species [*in andere Gattungen übergegangen sind*].[212]

Conclusion

With the *Biologie* of Treviranus the transcendental biology of the Göttingen School had concluded its formative period. In that work he had succeeded in pulling together all the various aspects of the program that had been under intense discussion since 1790, the individual elements of which can be traced back to discussions beginning in 1750 with the translation of Buffon's *Histoire naturelle*. Treviranus synthesized these conceptual elements into what he described as the dynamic theory of organic nature, which he attempted to ground in an incredible encyclopedic overview of biological research since the mid-eighteenth century. In this study I have attempted to reassemble those various elements by following their genesis and ultimate integration in the work of Treviranus and his efforts to found a new science, which he called "Biology."

The Göttingen program was merely coming into being during the period discussed here, however. It continued to be developed after Treviranus' epoch-making work, and indeed it was the principal framework of biological research in Germany well into the nineteenth century. A thorough understanding of the Göttingen program provides a background against which we can interpret some key conceptual elements of early nineteenth-century biology and in terms of which some of the major threads of empirical research in German biology become consistently interrelated. In concluding this study I would like to call attention to some directions in which the program was developed after 1803, the date of publication of the second volume of Treviranus' *Biologie*. Three areas of research stand out as having an especially significant role for carrying through to completion the program we have seen outlined here: sensory physiology, embryology, and the construction of general laws for the geographical distribution of forms.

Central to the theory of organization developed in the Göttingen program, especially the formulation given to it by Kielmeyer, was an understanding of the laws governing the "number of contacts" an organism has with its environment. For this problem research in sensory physiology became absolutely essential. Certainly the general emphasis on subjective experience during the Romantic era helped to make this research more fashionable, even providing princes, universities, and governments with motivation for funding it;

but research in this domain was clearly envisioned as essential to understanding the organization of nature as a whole. This point was made clear by Rudolphi in 1810.[213]

In a paper delivered before the Berlin Academy, Rudolphi claimed that while the external forms of organisms are modeled in large part in accordance with the external conditions of their environment, so that "species, genera, families, and perhaps even orders can be identified by external characters, classes can only be identified by internal principles of organization."[214] Rudolphi insisted that a system based on the internal principles of the entire organization of the animal ought to be the goal of physiology, but one system in particular serves as a key to that organizational plan: the nervous system. In a very important discussion of this problem, which has been overlooked by most historians, he came to the conclusion that there are four basic plans of neural organization in the animal kingdom and that while generative series can be constructed within each of these classes, there is no possibility of transforming one plan into another.[215]

It is within the framework of the Göttingen program that much of Johannes Müller's research in sensory physiology is to be understood, although he was certainly also deeply influenced by Goethe. Müller's theory of specific sense energies, however, has direct roots in Kielmeyer's dynamic theory of vital forces and especially in the conception of an "organ" so central to Kielmeyer's *Rede.* An "organ" for Kielmeyer was the material expression of a dynamic interaction of a purposively organized set of forces. The material constitution of an organ permitted it to have a graded variety of specific adaptive responses to external stimuli. Moreover an organ possessed its own internal activity. Its functional operation was not just a passive response to external stimuli but rather an active engagement with the external world occasioned by external impulses. In accordance with this view, the "appearances" generated in this interaction were completely conditioned by the internal structure and functional character of the organ itself. This very idea was explored in the context of sensory physiology in 1795 by Sömmering in a treatise called *Über das Organ der Seele,* a book, incidentally, that was dedicated to Kant and which Kant acknowledged as in fundamental agreement with his own views.[216] But this conception is also in fundamental agreement in almost every particular with the theory later developed by Müller on the specific sense energies.

A second area of research clearly indicated as central to the concerns of the Göttingen program is embryology. That embryology might offer a key to understanding the laws of purposive organization in nature was central to Blumenbach's thought. We have seen in the work of Kielmeyer an explicit statement of the importance of embryology in this regard and an intimation of the view that ontogeny recapitulates phylogeny. This idea was taken up and explored directly by another Blumenbach student, Johann Friedrich Meckel.[217] Von Baer's embryological work can also be seen in the framework of the Göttingen program. Several aspects of his work bear the marks of the system we have treated here. Foremost among these is the emphasis he placed on developmental plans of internal organization grounded in a *zweckmäßig* set of interconnected forces. His notion of different grades of expression of these structural forces and his related idea of a radial scheme of classification as corresponding to the natural system bears very strong resemblance to ideas we have seen treated by Kielmeyer and Treviranus.

Another area of research the Göttingen School perceived as central to its aims was the external laws governing the distribution of animal and plant forms. We have seen this as a principal theme in the work of Kielmeyer, Treviranus, and Humboldt. It was also regarded by Rudolphi to be an absolutely critical problem demanding research.[218] Contributions toward this aspect of the theory were made by this first generation of practioners of the Göttingen program, but the success of the Göttingen School in attacking this problem was best exemplified by the work of the Göttingen physiologist Carl Bergmann in his formulation of the rule relating size, distribution, and temperature.[219]

Our understanding of the biological theories of the Romantic period will not be complete until we come to grips ultimately with the traditions of speculative and metaphysical *Naturphilosophie.* It is my thesis that before this—in many ways much more difficult—task can be attempted it is first necessary to understand the biological tradition of transcendental *Naturphilosophie.* I am convinced that a careful reading of Schelling, Hegel, and even Ritter, Oken, and Carus will reveal that for almost the entirety of the empirical research upon which they based their approach to nature, these men were dependent upon research emanating from Göttingen. These other biological traditions attempted to go beyond the program of the Göttingen biologists to treat problems that were inaccessable with the approach of transcendental *Naturphilosophie,* particularly in the realm of social life.

Goethe's connections to Göttingen biologists, and to Blumenbach and Humboldt in particular, have been well established.[220] Through careful reading of Schelling's early works on *Naturphilosophie* in conjunction with the critical edition of his correspondence provided in the excellent edition of Horst Fuhrmanns, it is possible to show that while in Leipzig, in the period during which he devoted himself almost exclusively to acquiring background in natural science, Schelling concentrated on the works of the Göttingen School, particularly Lichtenberg, Blumenbach, and Kielmeyer. Through direct personal contact with C. H. Pfaff and Eschenmaier, Schelling gained an in depth knowledge of Kielmeyer's "Physik des Tierreichs."[221] Hegel's *Naturphilosophie* is equally indebted to the Göttingen School. Having, after 1806, rejected Schelling's increasingly speculative approach to nature as "the night in which all cows are black," he turned principally to Treviranus' *Biologie* for the biological sections of his own *Naturphilosophie.*

In addition to these personal ties and indebtedness to the Göttingen School for much scientific material, there were other intellectual ties between transcendental *Naturphilosophie* and speculative theories of nature. The concepts of *Einheit, Stufenfolge, Polarität, Metamorphose, Urtyp,* and *Analogie* have been described as distinguishing characteristic elements of this approach to nature.[222] We can see strong family resemblances to some of these ideas in the work of the Göttingen School. The notion of ideal types, for instance, was central to Göttingen thought on comparative anatomy. Similarly, in the work of Kielmeyer the dynamic interaction of vital forces, the expansion of one and the corresponding contraction of others, bears strong similarity to Schelling's notion of polarity, as well as the use made of that notion in the work of Nees von Essenbeck and Oken. The concept of metamorphosis central to speculative thought is closely parallel to the modification of an original ground plan that we have seen developed in the work of the Göttingen biologists. Finally, we have seen in Treviranus' discussion of the analogy between the infusoria and the phenomena connected with the emergence of all forms of life an example of the use of the concept of *Analogie* not at all unlike that found in the speculative tradition.

It would be a mistake to regard these approaches to nature as one and the same, however. There are indeed strong similarities in key concepts of these approaches, but there are major differences in both the interpretation and significance of these concepts within the transcendental and speculative schools. It is most important to realize that the transcendental approach worked hard at remaining consistent with Kant's philosophy of organic nature. The starting point for Schelling, Hegel, and Goethe, on the other hand, was made precisely in the conscious attempt to transcend the viewpoint of practical philosophy (i.e., moral and political philosophy) implied by Kant's critique of teleological judgment. The philosophical solution they worked out led them to construct a theory of nature that was much closer in the spirit of its conceptual foundations to the aesthetic conception of Buffon, namely, the approach the Göttingen School sought to avoid through strict adherence to empirical methodological canons. This preference for the aesthetic solution to the problems of the philosophy of biology can be seen reflected in the different notion of the *Urtyp* in the works of Oken, Goethe, and Carus. They seek to transform a particular shape or structure into a related set of forms. The Göttingen School, by contrast, emphasized the notion of a plan of functional organization. Their type, which they designated by the term *Grundform,* might be more appropriately labeled a "physiological type." Furthermore, it is important to note that Oken and Carus, for example, sought to construct a purely *deductive* and constitutive theory of organic nature, a theory that they grounded on a single unified force in nature. As we have seen in our discussion of both Kant and Treviranus, both of these aspects were considered by the Göttingen School to be a hyperphysical and illegitimate use of the principle of teleology. Schelling, too, emphasized the need for deducing all of nature from a single unity. Only in this way could the problem of distinguishing a collection of stones from a genuinely organized body encountered in Kant's theory of teleology be overcome. An organized body had to be constructed by differentiation of a single unity. The parts of this whole would not then be a collection of atomic organs held together somehow by the opposing force of an external nature; rather each part would only exist and have its life in the whole. In order to bring about the structural differentiation of nature Schelling argued that an original tension must exist in an originally homogeneous infinite material substrate, and that this tension must manifest itself as polarity, thereby providing the internal source for a structurally differentiated unity. Consonant with this view many speculative biologists, such as Oken, Goldfuß, and Carus, for example, made a direct analogy between nature as a whole and the mammalian ovum. This was not fashionable among Göttingen biologists, however, even though they too were committed strongly to an epigenetic theory of development. Thus Treviranus emphasized that the original organic matter was broken up into small organic spherelets, and that these became the building blocks for larger and more complex animals. He expressly denied that the unity of forces in nature could ever be grounded in a single unifying active principle. His definition of life always assumed the awakening of internal potential for reaction by means of an external nature.

In order to understand the philosophical issues behind the speculative tradition of biology more fully a trail will have to be blazed through some difficult passages in Schelling's early works, through Fichte's *Wissenschaftslehre* and through Hegel's *Logik*. We have seen that according to Kant the special nature of causal relations in the organic realm and the discursive nature of human understanding require that biology rest on regulative teleological principles. If one were to deny that thesis, however, as Hegel and Schelling did, and assert that the principle of teleological judgment can be *constitutive,* then in order to carry through the program of biological science consistent with this assumption, a new logic would have to be constructed, one in which constitutive determinate judgments can be rendered of objects that are both causes and effects of themselves. It was need of such a form of logic that could handle the special issues of organic casuality that led Schelling to attribute such a significance to polarity within his theory of organic nature, and it was this problem equally that led Hegel to formulate his own dialectical form of logic. Focusing on the special requirements of the theory of causality in biology will reveal the reason for the significance of other concepts in Romantic *Naturphilosophie*, such as *Urtyp*, *Stufenfolge*, etc., and it will reveal the structural interrelation of these concepts in a system of nature.[223]

In my view, however, it will be necessary ultimately to look beyond the substantive issues of these philosophical disputes to understand more fully the motivations of natural philosophers for preferring one over the other of these related approaches to biology in the Romantic era. In closing I will suggest one avenue for understanding this problem within a broader cultural perspective.

A thesis that has been widely accepted among German political historians is that the major political movements of the nineteenth century trace their source to a common fund of ideas in late Enlightenment thought. An exactly parallel phenomenon is evident in German biology. The main tenets of German liberal-conservatism, the view of society and the structure of the state to which they led, are reflected in and strongly supported by the view of nature developed by the Kantian school of biologists at Göttingen. The fact that the Göttingen legal faculty was a major center of liberal conservatism, particularly modeled after the English, may be indicative of the mutual support these two lines of thought were capable of lending one another; they were different aspects of the same world view. Similarly the biological theories of the metaphysical *Naturphilosophen* provided ample support for the conservative political ideology of the German Idealists and Romantics. In fact we can see from the first program statement of German idealism formulated by Schelling, Hegel, and Hölderlin, as well as from Hegel's early treatise on natural law that the originators of the speculative/metaphysical approach fashioned their view of organic nature much more consciously than did the Göttingen biologists in terms of a definite picture of certain social and political ends that they hoped to realize. The principle elements of the Volkish Ideology, which underpinned conservative culture throughout the nineteenth century,[224] are present in the biological theories of the Idealists; and the fact that Hegel was one of the major theoreticians of both conservative political theory and the Romantic-Idealist conception of nature cannot fail to alert us to the mutual affinity of these two aspects of German intellectual culture in the early nineteenth century.

Preference for one of these styles of *Naturphilosophie* rather than another, which differed more in terms of organizational emphases than in terms of empirical content, may reflect the response of German intellectuals to the main events affecting German society and culture at the turn of the century. Germany during the 1790s was entering a period of deep social and cultural crisis. The bonds of the feudal world were undergoing rapid dissolution, but within this general ferment, the outlines of the new world were not yet clearly visible. What was clear was that a return to the old world was impossible;

the effects in Germany of the French Revolution, the Napoleonic occupation, and the collapse of the Holy Roman Empire had permanently shut that door. While shouldering the burden of being thrust into the future, albeit in most cases with some reluctance, German intellectuals asked themselves whether it was necessary to sever completely the concrete ties to the past and rush forward headlong without any apparent direction, as the French had seemed to do, or whether it might be possible to preserve elements of the past, weaving them rationally into the fabric of a new state that might more appropriately realize the cultural aims of genuine freedom, morality, and the recognition of human dignity, which had been aborted in the misguided results of the Revolution.

These were the primary problems confronting German intellectuals in their daily lives; it simply was not possible to escape considering them. Various visions of how best to solve them and the practical implications for implementation of these solutions through political, social, economic, and educational reform provided the overarching framework of discourse, the set of socio-cultural givens within which other concerns took their meaning and orientation. Foremost among these was natural science, and the nascent life sciences in particular, for they purported to provide insight into man's place in nature and his ability to know and shape it. As heirs of the Enlightenment these men could not avoid the issue of whether nature was constituted in a manner consistent with the realization of human freedom and what the laws of nature, particularly the laws of organic nature, had to say about the organization of the state structured toward that end. It is not surprising, therefore, that the different styles of natural philosophy during the period 1790–1830, the differing views of nature and the organic realm, reflect in large measure the different political orientations of the period, for the sciences provided part of the rationale for constructing a particular vision of the future.

NOTES

1. C. C. Gillispie, *The Edge of Objectivity* (Princeton: Princeton University Press, 1960), p. 197.

2. Discoveries potentially attributable to the influence of Naturphilosophie have been electrochemistry (Ritter), electromagnetism (Oersted, Faraday), and the conservation of energy (Mayer, Helmholtz). On these problems see the following: Friedrich Klemm and Armin Hermann, eds., *Briefe eines romantischen Physikers* (Munich, 1966), p. 9, n. 2; Armin Hermann, "Das wissenschaftliche Weltbild Lichtenbergs," in *Aufklärung über Lichtenberg,* ed. Wolfgang Promies (Göttingen: Vandenhoeck and Ruprecht, 1974), p. 55; Walter D. Wetzels, *Johann Wilhelm Ritter: Physik im Wirkungsfeld der Deutschen Romantik* (Berlin-New York: De Gruyter, 1973), p. 25; R. C. Stauffer, "Speculation and Experiment in the Background of Oersted's Discovery of Electromagnetism," *Isis* 48 (1957): 33–50; L. Pearce Williams, *Michael Faraday* (New York: Basic Books, 1966), pp. 138 ff; T. S. Kuhn, "Energy Conservation as an Example of Simultaneous Discovery," in *Critical Problems in the History of Science,* ed. M. Clagett (Madison: University

of Wisconsin Press, 1955), pp. 321-56; Yehuda Elkana, *The Discovery of the Conservation of Energy* (Cambridge: Cambridge University Press, 1974), pp. 171-72, n. 31; Everett Mendelsohn, "Revolution and Reduction: The Sociology and Methodological and Philosophical Concerns in Nineteenth-Century Biology," in *The Interaction between Science and Philosophy,* ed. Yehuda Elkana (Atlantic Highlands, N.J.: Humanities Press, 1974), p. 419; Reinhard Löw, "The Progress of Organic Chemistry during the Period of German Romantic Naturphilosophie (1795-1825)," *Ambix* 27 (1980): 1-10. Löw has established the importance of Naturphilosophie for the development of structural chemistry; W. Riese, "The Impact of Romanticism on Experimental Method," *Studies in Romanticism* 2 (1962): 11-22.

3. Shmuel Sambursky, "Hegel's Philosophy of Nature," in Elkana, ed., *Interaction between Science and Philosophy,* p. 153.

4. For Novalis see Martin Dyck, *Novalis and Mathematics* (Chapel Hill: University of North Carolina Press, 1960); Kate Hamburger, "Novalis und die Mathematik; eine Studie zur Erkenntnistheorie der Romantik," *Romantik-Forschungen* (Hall, 1929), vol. 16; John Neubauer, *Bifocal Vision* (Chapel Hill: University of North Carolina Press, 1971). On Goethe see H. Bräuning-Oktavio, "Vom Zwischenkieferknochen zur Idee des Typus. Goethe als Naturforscher in den Jahren 1780-1786," *Nova Acta Leopoldina,* N.F. Nr. 126, vol. 18; Andreas Wachsmuth, *Geeinte Zwienatur* (Berlin, Weimar, 1966); H. B. Nisbet, *Goethe and the Scientific Tradition* (London, 1972).

On Oken and Carus see Rudolph Zaunick, "Oken, Carus, Goethe. Zur Geschichte des Gedankens der Wirbelmetamorphose," *Historische Studien und Skizzen zu Natur und Heilwissenschaft* (Berlin, 1930).

5. See especially Dietrich von Engelhart, *Hegel und die Chemie* (Wiesbaden: Guido Pressler, 1976). Also relevant is D. M. Knight, "The Physical Sciences and the Romantic Movement," *History of Science* 9 (1970): 54-75.

6. See Dietrich von Engelhart, "Naturphilosophie im Urteil der 'Heidelberger Jahrbücher der Literatur': 1808-1832," *Heidelberger Jahrbücher* 19 (1975): 35-82. For an interesting negative critique of Schelling Naturphilosophie see Kielmeyer's (anonymous) review in the *Tübinger Anzeigen* (Spring 1798). Other interesting critical reviews appeared in the *Allgemeine Literatur Zeitung* 4 (1799).

7. Thus Schelling could (mistakenly) argue that the approach to biology outlined by Kielmeyer in his famous lecture, "Über die Verhältnisse der organischen Krafte untereinander in der Reihe der verschiedenen Organisationen," (1793), was completely compatible with the system discussed in his own work, *Von der Weltseele.* See Schelling *Werke,* vol. 2 (1857), p. 565.

8. Kant, "Prolegomena zu einer jeden künftigen Metaphysik die als Wissenschaft wird auftreten können (1783)," *Werke,* vol. 4 (Berlin, 1903), p. 297.

9. Hegel, *Philosophy of Nature* (Oxford: Clarendon Press, 1970), tr. A. V. Miller, p. 37, section 246.

10. Reinhard Löw, *Die Planzenchemie zwischen Lavoisier und Liebig* (Munich: Donau Verlag, 1977); Reinhard Löw, *Die Philosophie des Lebendigen: Der Begriff des Organischen bei Kant, sein Grund und seine Aktualität* (Munich: Suhrkamp, 1980); see Dietrich von Engelhardt, *Hegel und die Chemie* (Wiesbaden: Guido Pressler Verlag, 1976). Also of interest is Peter Kaptza, *Die frühromantische Theorie der Mischung: Über den Zusammenhang von romantischer Dichtungstheorie und Zeitgenossischer Chemie* (Munich: Max Hueber Verlag, 1968). Kapitza makes the point that Romanticism not only affected the way the Naturphilosophen viewed chemistry but conversely that chemical theories of affinity also had a formative influence on the content of Romanticism itself. This is a marvelous study, which has received far too little attention in the literature; H.A.M. Snelders, "Romanticism and Naturphilosophie and the Inorganic Natural Sci-

ences 1797–1840: An Introductory Survey," *Studies in Romanticism* 9 (1970): 193–215; H.A.M. Snelders, "De outvangst van Kant bij enige Nederlandse Natuurwetenschaps beoefenaars omstreeks 1800," *Scientiarum Historia* 12 (1970): 22–38.

11. See Nelly Tsouyopoulos, "Die neue Auffassung der klinischen Medizin als Wissenschaft unter dem Einfluss der Philosophie im frühen 19. Jahrhundert," *Berichte zur Wissenschaftsgeschichte* 1 (1978): 87–100.

12. Georg Gruber, *Naturwissenschaftliche und medizinische Einrichtungen der jungen Georg-August-Universität zu Göttingen* (Berlin, 1955), p. 7.

13. Quoted from Johannes Joachim, Die Anfänge der Königlichen Societät der Wissenschaften zu Göttingen," *Abhandlungen der Gesellschaft der Wissenschaften zu Göttingen*, Philosophisch-Historische Classe, Dritte Folge, Nr. 19 (1936), 4.

The view being defended here that at Göttingen an interest in pure science was developing during the late eighteenth century differs sharply from the interpretation offered by R. Steven Turner, who places the development somewhat later. See R. Steven Turner, "The Growth of Professorial Research in Prussia, 1818 to 1848–Causes and Context," *Historical Studies in the Physical Sciences* (Baltimore: The Johns Hopkins University Press, 1972), 3, especially pp. 146–47 for Göttingen.

14. On this point see the excellent discussion in Roger Hahn, *The Anatomy of a Scientific Institution: The Paris Academy of Sciences 1666–1803* (Berkeley: University of California Press, 1971).

15. Extensive documentation for this point is provided in Johannes Joachim, loc. cit., fn. 1 above.

16. See Johan Stephan Pütter, *Versuch einer gelehrten akademischen Geschichte der Georg-August-Universität zu Göttingen,* I (Göttingen, 1765). A similar interest in theoretical matters was evident in many of the prize questions for the medical faculty as well. Thus the Preisfrage for 1766 ran: "Cum quasdam plantarum varietates credant botanici a diversorum generum commixtione (ut animalia hybrida) nasci: illam questionem experimentis, non conjecturis decidere; et si confirmetur veritas suspicionis huius, ad leges simul, quas sequunter istae varietates attendere." As we shall see this interest in finding the laws of variation among species became a central research problem for students in Göttingen in the late 1780s.

17. Gruber, loc. cit., p. 7.

18. On this point see the course descriptions included in Pütter, *Geschichte.*

19. On this point see Pütter, but also Christian Gottlob Heyne, *Akademische Vorlesungen über die Archäologie der Kunst des Altertums* (Braunschweig, 1822).

20. W. Brednow, *Jena und Göttingen* (Jena, 1949), p. 2.

21. Albrecht von Haller, "Vom Nutzen der Hypothesen" *Sammlung kleiner Hallerischer Schriften* (Bern, 1772), Teil I, p. 55.

22. Ibid., p. 55.

23. Ibid., p. 58.

24. Ibid., p. 59.

25. Ibid., p. 60.

26. Ibid., p. 67.

27. Buffon, *Allgemeine Historie der Natur* (Leipzig, 1750), tr. Abraham Gotthelf Kaestner, Bd. I., Teil I, p. 4. A similar statement is found in Bd. II, Teil I, p. 4. Because the notes to the German edition of Buffon's work prepared by Kaestner are useful in gaining an insight into the research plans of Göttingen naturalists during the 1760s and 1770s, I will cite the German text.

28. Ibid., p. 8.

29. Ibid., pp. 14–15.

30. Ibid., p. 26. I have translated "Arten" here by "genera," because Buffon clearly did believe in the physical reality of species. They were identified infallibly by the ability to produce fertile offspring.

31. Philip R. Sloan, "Buffon, German Biology and the Historical Interpretation of Biological Species," *British Journal for the History of Science* 12 (1979): 109–53.

32. Buffon, loc. cit., p. 8.

33. Ibid., p. 17.

34. Professor Sloan argues (loc. cit., fn. 31 above) that a tension existed between two approaches to natural history in late eighteenth-century German biological thought. There were those who defended a phenomenalistic approach, claiming that natural history could never be more than a description (*Naturbeschreibung*) of the phenomena. Opposed to this approach were those who claimed that a natural history could be constructed that got at the true interrelations among organisms. The proponents of this view argued that the historical succession of individuals would provide a system of nature. Thus they claimed that a genuine *Naturgeschichte* could be given. In Sloan's view, the defenders of this method were the true inheritors of Buffon's ideas. It seems to me that at Göttingen both of these approaches were pursued simultaneously, and both were seen to have been linked together as mutually supportive in Buffon's work. There may indeed have been a tension between the two methods, but it was one that Göttingen scientifists attempted to resolve.

35. This point will be made explicit below.

36. Buffon, loc. cit., Bd. II, Teil I, p. 16.

37. Ibid., p. 22. For a further discussion of this problem, see the excellent treatment by Paul Farber, "Buffon and the Concept of Species," *Journal of the History of Biology* 5 (1972): 259–84.

38. Cf. ibid., Bd. I. Teil I, pp. 12–13.

39. Ibid., Bd. II, Teil I, pp. 4–5.

40. See the summaries of Euler's correspondence concerning Büttner in *Euler Opera Omnia*, Series IV, Teil A, Bd. I. Register band, ed. A. P. Juschkewitsch (Basel, 1975).

41. Georg Ernst Stahl, Theorie der Heilkunde (Berlin, 1831). Translation of the *Theoria medica vera*, p. 5.

42. Ibid., p. 4.

43. Ibid., pp. 14–15.

44. Ibid., p. 12.

45. Johann Friedrich Blumenbach, *Beyträge zur Naturgeschichte* (Göttingen, 1805), p. 42.

46. Pütter, loc. cit., Teil I, pp. 290–91.

47. In his dissertation, *De generis humani varietate nativa* (Göttingen, 1775).

48. Pütter, loc. cit., p. 291.

49. Cf. Christian Wilhelm Bütter, *Vergleichungstafeln der Schriftarten verschiedener Völker in der vergangenen und gegenwärtigen Zeiten* (Göttingen, 1771–81), 2 vols.

50. Thus Buffon writes, loc. cit., Bd. IV, p. 102: "For every species in nature there exists a general type [*Urbild*] after which each individual is formed. The first animal, the first horse, is the external and internal model for all horses that have been and ever will be born. . . . But even though millions of individuals exist, not one of them has identical characteristics either to another individual or to the model." An obvious correlation existed in Buffon's theory between the *moule interieur* and the ideal type, underscoring once more the importance of generation for natural history. One crucial difference should be emphasized between Buffon's theory and that of the German naturalists, however. As the above passage indicates, the *Urbild* for Buffon was an in-

vidual horse or animal with an actual historical existence. All members of the same species were descendants of this original (pair) through the process of reproduction. For the German naturalists, however, the *Urbild* never had a concrete existence. Always subject to the effects of the environment, the *Urbild* itself never appeared in nature, rather only its phenomenal manifestations. On this point see the discussion on Girtanner and Blumenbach below as well as my paper, "Generational Factors in the Origin of *romantische Naturphilosophie*," *Journal of the History of Biology* 2 (1978): 57–100.

51. The connection between Heyne's views and those of Diderot was noted by Blumenbach in his *Knochenlehre* (2. Auflage, Göttingen, 1807), pp. 84–86.

52. Heyne, *Akademische Vorlesungen über Archaeologie* (Braunschweig, 1882), p. 13.

53. See Denis Diderot, "Essais sur la peinture," in *Oeuvres esthetiques* (Paris: Clasiques Garnier, 1959), p. 666. Compare with this Buffon, loc. cit., Bd. II, pp. 22–24. Kaestner points out in his note to this passage that the manner in which reason affects this penetration to the "unsichtbar innerliche Form" is not at all clear.

54. Götz von Selle, *Die Matrikel der Georg-Augustus Universität zu Göttingen* (Hildesheim, Leipzig, 1937), Bd. 1, p. 198.

55. Blumenbach, *Treatise on Man*, pp. 97–99. All selections from the *De generis varietate humani nativa* quoted in this paper have been taken from *The Anthropological Treatises of Johann Friedrich Blumenbach* (London, 1865) tr. Thomas Bendyshe.

56. Ibid., p. 188.

57. Ibid., pp. 188–89.

58. Ibid., p. 190.

59. Ibid., pp. 190–91.

60. Ibid., pp. 74–76.

61. Ibid., p. 71.

62. Blumenbach, *Handbuch der Naturgeschichte* (Göttingen, 1802, sixth ed.), p. 8.

63. Blumenbach, *Handbuch der vergleichenden Anatomie* (Göttingen, 1805), pp. v–vi.

64. Caspar Friedrich Wolff, "De inconstantia fabrica corporus humani," *Acta Petropolitianae* (1778), p. 218.

65. Ibid., p. 226.

66. Diderot, "Essais sur la peinture," in *Oeuvres esthetique* (Paris: Classiques Garnier, 1959), pp. 665–66.

67. Blumenbach, *Handbuch der vergleichenden Anatomie*, pp. 84–86.

68. This is catalogued in the Blumenbach Nachlass at Göttingen as Cod. Ms. Blumenbach XXIX, über *Monstra*.

69. Cod. Ms. Blumenbach XI, in the unnumbered handwritten introductory material to the *Handbuch der vergleichenden Anatomie* used by Blumenbach for his lectures.

70. Buffon, *Allgemeine Historie der Natur*, Bd. II, pp. 23–24.

71. That Haller and Blumenbach corresponded on this problem was acknowledged explicitly by Blumenbach in his treatise, *Über den Bildungstrieb* (Göttingen, 1781), p. 6.

72. Haller, "De partibus corporus humani sensibilibus et irritabilibus," *Commentarii Societatis Regiae Scientarum Göttingensis*, Tom II (1753), pp. 114–58. See as well Haller's statement of his problem in his extremely popular *Grundriss der Physiologie* (Berlin, 1788), Meckel, Wrisberg and Sömmering (eds.), pp. 69–70 and p. 311.

73. Blumenbach, *Handbuch der Naturgeschichte*, p. 19.

74. Ibid., p. 19.

75. Blumenbach, *Über den Bildungstrieb*, p. 6.

76. Ibid., p. 10.

77. Ibid., pp. 12–13.

78. Ibid., p. 19.

79. Ibid., p. 14.

80. Ibid., pp. 55–56.

81. Cod. Ms. Blumenbach XII, *Handbuch der Naturgeschichte* (1799), pp. 615–16.

82. Blumenbach, *Beyträge zur Naturgeschichte,* pp. 6–8.

83. Ibid., p. 19.

84. Ibid., p. 20.

85. Ibid., pp. 19–20. On p. 25 Blumenbach asserts: "And just as nothing prevents a species from being destroyed, so on the other hand a new one can be created [*nacherschaffen*] from time to time."

86. Ibid., p. 23.

87. Blumenbach, *Über den Bildungstrieb,* p. 43.

88. Ibid., p. 42.

89. Ibid., p. 62.

90. Ibid., p. 61.

91. Blumenbach, *Beschreibung und Geschichte der Knochen des Menschlichen Korpers* (Göttingen, 1807, second ed.), p. 41.

92. Ibid., p. 41.

93. Ibid., p. 6.

94. Haller, *Grundriss der Physiologie,* p. 4.

95. Ibid., p. 4.

96. Ibid., pp. 8–9.

97. Blumenbach, *Beschreibung und Geschichte der Knochen des Menschlichen Korpers,* p. 10.

98. Kant, "Die Bestimmung des Begriffs einer Menschenrasse," *Berlinischer Monatschrift* (1785), in *Kant's Gesammelte Schriften,* vol. VII. Kant, "Über den Gebrauch teleologischer Prinzipien in der Philosophie," Teutscher Merkur (1788), in *Kant's Gesammelte Schriften,* vol. VIII.

99. See Phillip R. Sloan, "Buffon, German Biology, and the Historical Interpretation of Biological Species," *British Journal for the History of Science* 12 (1979): 109–53.

100. Timothy Lenoir, "Kant, Blumenbach, and Vital Materialism in German Biology," *Isis* 71 (1980): 77–108.

101. Kant, *Gesammelte Schriften,* vol. XI, p. 176.

102. Kant, *Kritik der Urteilschraft* in *Kant's Gesammelte Schriften,* ed., Königlichen Preussischen Akademie der Wissenschaften (Georg Reimer: Berlin, 1902–23), vol. V (1908), p. 373.

103. Ibid., p. 371.

104. Ibid., p. 360.

105. Ibid., p. 179.

106. Ibid., p. 182.

107. Ibid., p. 375.

108. Ibid., p. 424.

109. Kant, "Über den Gebrauch teleologischer Prinzipien in der Philosophie," *Gesammelte Schriften,* vol. 8, p. 179.

110. For a fuller discussion of this point see my article, "Kant, Blumenbach and Vital Materialism."

111. Kant, *Kritik der Urteilskraft,* in *Gesammelte Schriften,* Vol. 5, p. 419.

112. Ibid., p. 418. The approach advocated by Kant, Blumenbach, and, as we shall see, Kielmeyer has many points of close similarity to ideas being developed contemporaneously by Cuvier. He too advocated a teleological approach based on plans of organiza-

tion to the fundamental unit. There were significant differences, however. Cuvier denied the environmental elements of the approach being developed by the Göttingen School. His theory was rabidly ahistorical and anti-developmental as evidenced by his rejection of Lamarck's ideas and his neglect of embryology, which came to be a central feature of the later work of the Göttingen traditions. On this issue see William Coleman, *George Cuvier, Zoologist: A Study of the History of Evolution Theory* (Cambridge: Harvard University Press, 1964), especially Chapters 2 and 3 and p. 157 ff. This problem is treated in detail in my forthcoming book, *Vital Materialism in Nineteenth Century German Biology.*

113. Ibid., p. 419.

114. For Blumenbach's path to a reconciliation of his views with this aspect of Kant's position, see my article in *Isis,* loc. cit.

115. Shirley Roe, "Rationalism and Embryology: Caspar Friedrich Wolff's Theory of Epigenesis," *Journal of the History of Biology* 12 (1979): 1–43.

116. Kant, *Kritik der Urteilskraft,* p. 419. On the difference between the functional theory proposed by Kant and the Göttingen school and the descent theory of Darwin, see the important paper by William Coleman, "Morphology between Type Concept and Descent Theory," *Journal of the History of Medicine* 31 (1976): 149-75. My own views on this problem are deeply indebted to Professor Coleman's work.

117. Ibid., p. 420.

118. For a discussion of the relationship between Kant, Girtanner and Blumenbach, see Phillip R. Sloan, loc. cit. in note 31 above, pp. 137-38.

119. Christian Heinrich Pfaff, *Über tierische Elektrizität und Reizbarkeit* (Leipzig, 1795), pp. 2-3, 234-36.

120. Christian Heinrich Pfaff, *George Cuviers Briefe an C. H. Pfaff aus den Jahren 1788-1792. Nebst eine biographischen Notiz von Pfaff* (Kiel, 1845). See especially February 1790, p. 141, February 19, 1791, p. 213, and August 1792, pp. 283–84.

121. Cf. Reinhard Löw, *Die Pflanzenchemie zwischen Lavoisier und Liebig* (Straubing and Munich, 1977), particularly Chapter 3.

122. Blumenbach, *Handbuch der Naturgeschichte* (1797), p. 24n.

123. Ibid., p. 24.

124. Kielmeyer, "Entwurf zu einer vergleichenden Zoologie," in *Gesammelte Schriften,* ed. F. H. Holler (Berlin, 1938), pp. 17-19. The importance of this organ for the classification of animals was emphasized by Blumenbach as well as later by Meckel, Agassiz, and Johannes Müller.

125. Ibid., pp. 25-26.

126. Ibid., p. 26.

127. Ibid., p. 27.

128. Ibid., pp. 28-29.

129. Kielmeyer, "Ideen zu einer allgemeineren Geschichte und Theorie der Entwicklungserscheinungen der Organizationen," in *Schriften,* p. 107. In a Kantian vein he goes on to indicate the importance of temporal relations for indicating lawlike patterns of phenomena as our only possible sign of the organization of things in themselves.

130. Ibid., pp. 122-23.

131. Ibid., p. 123.

132. Kielmeyer, "Ideen zu einer Entwickelungsgeschichte der Erde und ihrer Organizationen" in loc. cit., pp. 205-6.

133. Ibid., pp. 207-8.

134. Cf. Blumenbach, *Handbuch der Naturgeschichte* (1802), pp. 8-9.

135. Kielmeyer, loc. cit., note 129 above, p. 209.

136. Ibid., p. 210.

137. Ibid., p. 209.

138. For further discussion of this point Cf. William Coleman, "Limits of the Recapitulation Theory: Carl Friedrich Kielmeyer's Critique of the Presumed Parallelism of Earth History, Ontogeny and the Present Order of Organisms," *Isis* 64 (1973): 341-50.

139. Carl Friedrich Kielmeyer, "Über die Verhältniße der organischen Kräfte untereinander in der Reihe der verschiedenen Organization: Die Gesetze und Folgen dieser Verhältniße," in *Gesammelte Schriften*, p. 63.

140. Ibid., pp. 66-67.

141. Ibid., p. 67.

142. Ibid., p. 71.

143. Ibid., p. 75.

144. Ibid., p. 77.

145. Ibid., p. 78.

146. Ibid., p. 80.

147. Ibid., p. 85.

148. Ibid., p. 86.

149. Ibid., p. 89.

150. Ibid., p. 89.

151. Ibid., pp. 91-92.

152. Ibid., p. 98.

153. Alexander von Humboldt, *Aphorismen aus der chemischen Physiologie der Pflanzen* (Leipzig, 1794), p. 9. This was a German translation of Humboldt's Latin text by Gotthelf Fischer.

154. Ibid., p. 5.

155. Ibid., pp. v-vi.

156. Humboldt, *Ideen zu einer Physiognomik der Gewächse* (Stuttgart, 1871, 3rd ed.), Bd. 2, p. 16.

157. Ibid., pp. 21-22.

158. Ibid., pp. 24-25.

159. Cf. ibid., p. 121 ff.

160. Ibid., pp. 124-25. See also Alexander von Humboldt, *De distributione geographica plantarum secundum coeli temperiem et altitudinem montium* (1817), pp. 24-44.

161. Ibid., pp. 122-24.

162. Kosmos Humboldt, *Entwurf einer physischen Weltbeschreibung* (Stuttgart, 1845), Bd. I, pp. 55-56.

163. Ibid., pp. 63-64.

164. Ibid., p. 63.

165. Humboldt, *Ideen zu einer Physiognomik der Gewächse*, p. 15. In discussing this point Humboldt cites Heinrich Friedrich Link, with whom he had also studied at Göttingen. The passage quoted from Link is as follows: "In the *Urwelt* we find the most distant structures pressed together in the most amazing forms, pointing at the same time to great development and division [*Entwicklung und Gliederung*] in the *Nachwelt*." Quoted from *Abhandlungen der Akademie der Wissenschaften zu Berlin,* Jahrgang 1846, p. 322. Link expressed similar views in other works as well. Cf. his *Natur und Philosophie* (Rostock und Leipzig, 1811), pp. 336-41. An almost identical set of views is set forth by G. R. Treviranus in his *Biologie.*

166. Gottfried Reinhold Treviranus, *Biologie: Oder Philosophie der lebenden Natur* (Göttingen, 1802-22), 1: 4.

167. Ibid., p. 17.

168. Ibid., pp. 32-34.
169. Ibid., pp. 35 and 55-56.
170. Ibid., p. 50.
171. Ibid., p. 34.
172. Ibid., p. 65.
173. Ibid., p. 69.
174. Ibid., pp. 69-70.
175. Ibid., pp. 98-99.
176. Ibid., p. 98.
177. Ibid., pp. 160-61.
178. Ibid., pp. 162-63.
179. Ibid., pp. 165-66.
180. This problem is examined by Reinhold Löw, *Philosophie des Lebendigen. Der Begriff des Organischen bei Kant, sein Grund und seine Aktualität* (Munich: Suhrkamp Verlag, 1980), pp. 227-29, pp, 284-308. Löw's work not only provides the key to understanding the philosophical foundations of the life sciences in the Romantic era, it is also extremely relevant to problems in contemporary biology and organic chemistry.
181. Treviranus, loc. cit., pp. 175-76.
182. The principal works cited by Treviranus in this section of his *Biologie* are Blumenbach, *Naturgeschichte,* Linnaeus, Georg Forster, Haller, Buffon, and Cuvier, *Vorlesungen über die vergleichenden Anatomie* (1801).
183. Ibid., pp. 179-80.
184. Ibid., pp. 180-81.
185. Ibid., p. 185.
186. Ibid., p. 192.
187. Ibid., p. 198.
188. Ibid., p. 235.
189. Ibid., p. 258.
190. Ibid., p. 475.
191. Ibid., p. 470.
192. Treviranus, *Biologie,* vol. 2, p. 157.
193. Ibid., p. 157.
194. Ibid., pp. 171-72, and p. 191.
195. Ibid., p. 135; see also vol. 3, p. 8 ff.
196. Ibid., p. 126.
197. Ibid., p. 173.
198. Ibid., p. 205.
199. Ibid., p. 85.
200. Ibid., p. 264.
201. See Elizabeth Gasking, *Investigations into Generation: 1651-1828* (London: Hutchinson, 1967) and Jacques Roger, *Les sciences de la vie dans la pensée français du xviiie siècle* (Paris: Armand Colin, 1963).
202. Ibid., p. 297.
203. Ibid., p. 269.
204. Ibid., pp. 271-77.
205. Ibid., p. 325.
206. Cf. vol. 1, p. 458.
207. Ibid., p. 377.
208. Ibid., p. 378.
209. Ibid., pp. 400-403.

210. Ibid., p. 407.

211. Ibid., p. 453.

212. Vol. 3, pp. 225–26.

213. Rudolphi was closely connected to Göttingen, having worked with Link on a project in 1805 that won the prize of the Göttingen Societät der Wissenschaften. These two were later colleagues at the University of Berlin. Moreover, Rudolphi expressly states his full agreement with Treviranus' approach in the *Biologie*; cf. Rudolphi, *Beiträge zur Anthropologie und allegemeine Naturgeschichte* (Berlin, 1812), p. 131.

214. Rudolphi, "Über eine neue Einteilung der Tiere," in *Beyträge*, p. 83.

215. See especially pp. 95–106.

216. See Kant's letter to Sömmering, *Kant's Gesammelte Schriften,* vol. 12, pp. 30–35.

217. See Stephen Gould, *Ontogeny and Phylogeny* (Cambridge: Harvard University Press, 1977), pp. 46–47. Meckel's work is treated in some detail in my *Vital Materialism in Nineteenth Century Biology* (forthcoming).

218. Rudolphi, loc. cit., pp. 109–72 in his lecture entitled "Über die Verbreitung der organischen Körper."

219. See William Coleman, "Bergmann's Rule: Animal Heat as a Biological Phenomenon," *Studies in the History of Biology* (Baltimore: The Johns Hopkins University Press, 1979), 3: 67–88.

220. See H. Bräuning-Oktavio, "Vom Zwischenkieferknochen zur Idee des Typus: Goethe als Naturforscher in den Jahren 1780–86," *Nova Acta Leopoldina,* N. F., Nr. 126, vol. 18; F. T. Bratranek, *Goethes Briefwechsel mit den Gebrüdern von Humboldt: 1795-1832* (Leipzig, 1876); W. Brednow, *Jena und Göttingen: Medizinische Beziehungen im 18. und 19. Jahrhundert* (Jena, 1949); Georg Uschmann, *Der morphologische Vervollkommungsbegriff bei Goethe und seine problemeschichtliche Zusammenhänge* (Jena, 1939).

221. See Horst Fuhrmanns, ed., *Schelling Briefe* (Bonn: H. Bouvier, 1973).

222. See Reinhard Löw, *Die Pflanzenchemie zwischen Lavoisier und Liebig* (Munich: Donau Verlag, 1977); Brigitte Hoppe, "Polarität, Stufung und Metamorphose in der spekulative Biologie der Romantik," *Naturwissenschaftliche Rundschau* 20 (1967): 380–83.

223. This problem is treated in the forthcoming book by Reinhard Löw, *Das teleologische Denken.*

224. See, for example, Fritz Ringer, *The German Mandarins* (Cambridge: Harvard University Press, 1968); George Mosse, *The Crisis of German Ideology* (New York: Grosset and Dunlap, 1964).